Food Processing

Edited by Romina Alina Marc,
Antonio Valero Díaz
and Guiomar Denisse Posada Izquierdo

Published in London, United Kingdom

IntechOpen

Supporting open minds since 2005

Food Processing
http://dx.doi.org/10.5772/intechopen.82985
Edited by Romina Alina Marc, Antonio Valero Díaz and Guiomar Denisse Posada Izquierdo

Contributors
Nícolas Mazzola, Claire Sarantopoulos, Michelle Guimarães Horta, Poliana Mendes De Souza, Fabiana Regina Lima, Carlos Alberto Gois Suzart, Cedomir Radovic, Milica Petrović, Marija Gogić, Dragan Radojković, Vladimir Živković, Nenad Stojiljković, Radomir Savic, Cristiane Marangoni, Alexandre Cichoski, Juliano Barin, Maged E. A. Mohammed, Mohammed Alhajhoj, Rotondi Annalisa, Lucia Morrone, Francesca Rapparini, Gianpaolo Bertazza, Rosa María García-Gimeno, Guiomar Denisse Posada-Izquierdo, Michael Nickerson, Roberto Rodríguez-Rámírez, Blanca Areli Mondaca-Navarro, Alma Guadalupe Villa-Lerma, Luz Angelica Avila-Villa, Gabriel Davidov-Pardo, Zohra Benaouf, Suzana Caetano Da Silva Lannes, Maria Elena Bernal Gómez, Romina Alina Marc

First published in London, United Kingdom, 2020 by IntechOpen
IntechOpen is the global imprint of INTECHOPEN LIMITED, registered in England and Wales, registration number: 11086078, 7th floor, 10 Lower Thames Street, London, EC3R 6AF, United Kingdom
Printed in Croatia

British Library Cataloguing-in-Publication Data
A catalogue record for this book is available from the British Library

Additional hard and PDF copies can be obtained from orders@intechopen.com

Food Processing
Edited by Romina Alina Marc, Antonio Valero Díaz and Guiomar Denisse Posada Izquierdo
p. cm.
Print ISBN 978-1-78985-893-8
Online ISBN 978-1-78985-894-5
eBook (PDF) ISBN 978-1-83880-375-9

For EU product safety concerns:
IN TECH d.o.o., Prolaz Marije Krucifikse Kozulić 3, 51000 Rijeka, Croatia, info@intechopen.com or visit our website at intechopen.com.

Meet the editors

Romina Alina Marc completed her PhD (Agronomy) in 2015 at the University of Agricultural Sciences and Veterinary Medicine (UASVM), Cluj-Napoca (Romania). She has a PhD Eng. and is a lecturer responsible for research activity in vegetable food quality control, rheology in the food industry, quality management systems, and food safety at the Faculty of Food Science and Technology, UASVM, Cluj-Napoca. She has published 36 research articles in reputed journals on the development of innovative food products and bioactive compound traceability during processing. She was responsible for seven research projects and has won two national awards and 42 international awards.

Prof. Antonio Valero Díaz holds a degree in Biology and a PhD in Food Science and Technology at the University of Cordoba. He has currently more than 10 years' experience managing projects, training courses, academic teaching, and events in food and related sectors. Antonio Valero has participated in different national and international research projects (>15) related to predictive modeling and risk assessment. His research activity is widely shown as he has published more than 70 peer-reviewed papers and book chapters mainly related to the application of predictive models in foods, microbial risk assessment and management, and derivation of microbiological criteria in selected foods. Additionally, he has presented more than 80 communications in different congresses and symposia since 2003.

Prof. Guiomar Denisse Posada Izquierdo completed her degree in Agronomic Engineering in 2006, her agrifood master in 2007, and received her PhD in Food Science and Technology from the University of Córdoba, Spain, in 2013, where she currently works in the Department of Food Science and Technology. She has more than 12 years' experience participating in the food quality and safety field together with training courses, academic teaching, and participation in food-related events. Guiomar Denisse Posada has been involved in different national and international research projects related to predictive modeling and risk assessment (development of a common structure for microbiological risk assessment and management in minimally processed and ready-to-eat foods: application in vegetables products, and selection of fit-for-purpose sampling procedures for specific foods and risks, BASELINE, COST Action 920). She has been trained and specializes in disinfection treatments and processing control of vegetable foods thanks to different pre- and post-doctoral stays at prestigious institutions such as the Department of Food Science and Human Nutrition, Michigan State University (USA), University of Ghent (Belgium), and the Spanish Research Council (CEBAS-CSIC and IG-CSIC). Her publications include more than 30 peer-reviewed papers and book chapters related to the application of predictive models in foods, microbial risk assessment and management, decontamination technologies in vegetables, and food shelf-life. Additionally, she has presented more than 50 communications in different conferences and symposia since 2006. She participated in the organizing committee of the last ICPMF10 conference held in Córdoba (September 2017). Currently, Guiomar Denisse Posada holds a position at the Department of Food Science and Technology (University of Córdoba). She has wide experience (8+ years) in academic teaching within the food science and technology and veterinary degrees.

Contents

Preface

Nowadays, consumers tend to request minimally processed additive-free food, with an extended shelf-life. Consumers' demands have exceeded the food taste barrier, with more and more nutritious food requirements that can meet nutritional needs.

Food processing is one of the key factors that diminishes the nutritional properties of food products alongside the transition from the raw to the processed state.

Numerous ongoing research studies are carried out into the use of minimal thermal methods to provide high-quality food.

Over the last two decades in response to consumer expectations, researchers in the food industry and government institutions have examined and explored the development of functional food products and new techniques for processing, preserving, and controlling food products.

A functional food product is a product that is supposed to have a supplementary function achieved by the addition of one or more new ingredients, with beneficial effects on people's health. To meet consumers' demands with regard to this type of product and to produce these foods, the processing of such food products is in continuous development and innovation.

In most cases, fresh raw materials are negatively affected within the processing procedure.

A major challenge in the food industry is to preserve the natural functional properties and the sensory and nutritional quality of the raw materials, including the appropriate shelf-life and safety. In addition, the potential for microbiological contamination of foods is high due to the conditions in the food traceability chain being extremely diverse.

Trends for minimally processed foods come with three approaches, as follows: the first approach is the optimization of traditional conservation methods for sensory improvement, nutritional and microbiological quality of food, and energy efficiency.

The second approach refers to the development of minimally invasive processes through new preservation combinations to obtain quality products regarding the fresh state of the given food, but with a longer shelf-life.

The third approach refers to the development of innovative techniques for obtaining foods of superior nutritional quality by using emerging preservation factors (e.g. high-pressure processing, pulsed electric field processing, cold plasma treatment, ultrasound processing, irradiation, UV, and pulsed light).

Food Processing aims to present new trends in the current field of science knowledge. The chapters in this book have been divided into four sections. Before the presentation of these specific sections, an introductory chapter proposes an overview of the most recent general information on new trends in food processing.

Section 1—Food Technologies in Food Processing—presents current technological processes used in food processing. Section 2—Quality of Raw Materials in Food Processing—presents the importance of the quality of raw materials used in food processing. Section 3—Treatments Used in Food Processing—presents the latest trends in treatments used in food processing. Section 4—Factors That Influence Food Processing—presents current information on the factors that influence food processing from the raw material to the packaging used.

My special thanks go to Ms. Katarina Pausic and Ms. Marijana Francetic, the Author Service Managers, for their support provided during the whole process of publication. I would also like to thank the authors who contributed chapters to this book. My thanks also go to the technical publisher who professionally prepared this book for publication by IntechOpen. I also want to thank my colleagues at the University of Agricultural Sciences and Veterinary Medicine, Cluj-Napoca, who supported me throughout this entire process, and last but not least my family for their whole-hearted support.

Romina Alina Marc
Faculty of Food Science and Technology,
University of Agricultural Sciences and Veterinary Medicine Cluj-Napoca,
Romania

Antonio Valero Díaz and Guiomar Denisse Posada Izquierdo
Department of Food Science and Technology,
University of Córdoba,
International Campus of Excellence in the AgriFood Sector (CeiA3),
Córdoba, Spain

Section 1

Introduction

Chapter 1

Introductory Chapter: A Global Presentation on Trends in Food Processing

Romina Alina Marc

1. Introduction

Nowadays food processing has reached a level that is hard to imagine. Food processing is almost omnipresent. Almost all food consumed in almost all settings is now processed in some way. Varied types of food processing have beneficial or even negative effects on food, diet quality, and human health [1].

A range of new food processing technologies has been investigated and developed to modify or replace traditional food processing techniques so that better quality and more consumer preference-oriented foods can be manufactured [2]. The focus has been on quality over the last decade, enhancing the process efficiency, safety, productivity, and stability of food products in a healthier way [3].

The nutritional quality of food is influenced by factors such as quality of raw material, transportation, processing techniques, packaging, storage, and the whole food chain (farm to fork) [4, 5].

Several existing methods and techniques included in food processing are used in order to transform raw material ingredients into food for people's consumption. Food processing implies the use of clean, butchered, and harvested components to manufacture food products for the market demand. There are numerous ways to produce food [6, 7].

Nowadays, an impressive displacement process is taking place worldwide within alimentary production patterns and systems, with regard to the ready-for-consumption food, which replaces traditional food based on fresh meals. However, not much attention is being paid to food processing, when it comes to dietary guidelines, classifications of food products, or epidemiology studies [8].

Scientists in this industry regard food processing in various ways. Certain processes such as drying, nonalcoholic fermenting, skimming, pasteurization, freezing, and vacuum-packing are rightly considered as beneficial in the food industry. However, other processes are regarded as less beneficial or even harmful for human health, namely, charring, hydrogenation, carbon dioxide addition, alcoholic fermenting, salt-pickling, and sugaring [1, 7].

Nowadays, a drastic change has been noticed in how compliance and its consequences are approached. Due to regulatory requirements, and to a much less tolerant and most knowledgeable public, the industry has been in the critical public eye and placed under a much more thorough scrutiny and examination than ever before. As a consequence, at the local and state levels, there has been an increasing demand of documentation, in terms of legislation. It is very difficult to make a mistake and have no one find out about it. In the end, we are actually all consumers

IntechOpen

and expect safe food. Legislation is seeing a lot of issues related to allergens, organics, and package labeling [9–11].

Food safety technology has greatly interfered with and changed the way food industry specialists identify, react to, and communicate on specific food safety issues. Thus, the Internet and social media have impacted the industry by allowing an instant information exchange, which, at times, can include true or even false data, hence rightly entitling food companies to have a correspondingly adequate reaction. The legislation regarding the food industry attempts to regulate the management of pests, sanitation, and contamination with fewer toxic materials. Having said that, there will always be people trying to "beat" any law or system; consequently it is extremely important to plan for that by being able to identify issues with strong systems and processes. The trap, however, lies in believing that laws or government can control food safety. For managing the hazards involved and the control over food safety, the legislation needs to constantly improve alongside with its enforcement, according to the Hazard Analysis Critical Control Points (HACCP) model [12, 13].

Services, processes, or innovations, understood as new products, are recognized as an important instrument for companies belonging to the food industry to stand out from competitors and to satisfy consumer expectations [14, 15].

Notwithstanding, nowadays, consumers' preferences for food have changed significantly; in fact, consumers believe that food should directly contribute to their health [15, 16].

In these circumstances, the demands of consumers are no longer limited to satisfying hunger. Food should provide the necessary nutrients while at the same time preventing nutrition-related diseases and improving physical and mental well-being [17].

Recent food industry innovations mainly regard new technical and scientific approaches in food processing, as well as the implementation of new food products on the market. In this respect, due to the constantly rising cost of healthcare, the steady increase in life expectancy, and the necessity for the elderly to enjoy an improved life quality, functional food plays a major role in the consumers' demand nowadays. As such, researchers agree in stating that functional food represents one of the most rewarding and gratifying areas of research and innovation in the food industry [18, 19].

The food industry is a key branch in the global economy. As a consequence, many authors highlighted its relevance for employment and economic output [9, 11, 14].

2. New trends in food processing

Nowadays, a major trend in food manufacturing has been influenced by the consumers' demand of functional or health-promoting foods. According to the Food and Nutrition Board of the Conform Institute of Medicine, functional foods are defined as "any food or food ingredient that may provide a health benefit beyond the traditional nutrients it contains." The term "functional food" was first used in Japan in the mid-1980s, and it refers to processed nutritious foods containing supplementary ingredients that give support to specific functions of the human body [1, 7, 20].

In the year 2000, Sloan presented another option for the definition of functional food, namely, "a food or beverage that imparts a physiological benefit that enhances overall health, helps prevent or treat a disease/condition, or improves physical or mental performance via an added functional ingredient, processing modification, or biotechnology" [21, 22].

Señorans et al. in 2006 classified health-promoting foods into three different categories, according to the types of food processing and their influence on human health, as follows: (1) those with specific functionalities, (2) foods fortified with natural ingredients providing a wanted functionality (foods enriched with natural ingredients), and (3) probiotics and prebiotics [7]. This classification is topical nowadays, but it is constantly innovating.

2.1 Functional foods

The major processes based on biotechnology are mainly used in the production of foods with specific functionalities. In this respect, genetic engineering constitutes an important support. Consequently, the industry's final objective of obtaining new foods of specific composition, with new and better functional properties, can be achieved, as these new biotechnologies can alter the genes contained in certain cells [23, 24].

The raw materials with modified composition constitute another promising area for producing foods with enhanced nutritional value. For example, foods having improved nutritional value began to appear 20 years ago, such as tomatoes with higher lycopene content or corn with higher oleic acid content [25] and white wines with a higher concentration of resveratrol [26], hypoallergenic foods, in which a specific protein or peptide has been removed [27], etc.

Functional food products have also been produced by the wide use of enzymes. Hence, new advances in this area have been used/are being used such as techniques of enzyme and cell immobilization or new developments in bioreactors. For example, milk protein concentrate with a decreased content of lactose [28] and fats with a controlled content of fatty acids [29] can be produced using the abovementioned techniques.

Another important technique is the application of membrane technology, which is used in modifying the composition of foods and their functionality [30]. This process lends itself mostly to foods in a liquid medium. For example, this method has proven its usefulness by allowing separation and concentration of milk components without the proteins, bioactive substances, and flavor being altered [31]. In addition, various processes have been developed in the dairy industry, for instance, ultrafiltration [32], nanofiltration [33], electrodialysis [34], and lactoferrin [35].

Another technological process is the removal of antinutrients by means of supercritical fluid extraction. Antinutrients are compounds which are thought of as negative or non-healthy ones, such as fats, caffeine, cholesterol, etc. [36]. This technology has been mainly used since the 1990s, for the elimination of caffeine from coffee [37] (this improved technique is still used today [38]), alcohol from cider [39] and wine [40], and fat from foods such as French fries, onion rings, and snack foods [41]. This improved technique is used today for extraction of raspberry seed oil [42], extraction of oleoresins and plant phenolics [43], bioactive compounds from plants [44], and food quality and food safety evaluation [45].

2.2 Foods enriched with natural ingredients

The addition of iron to bread or vitamin A and D to milk is an old and well-known procedure called "food fortification." The loss of nutrients while food is being processed has been overcome lately by supplementing health-promoting ingredients or nutraceutical ingredients to food products [24, 25].

The most important elements used for food fortification are iron, vitamin A, iodine, folate (vitamin B9), vitamin B12, other B vitamins (thiamine, riboflavin, niacin, and vitamin B6), vitamin C, vitamin D, calcium, selenium, fibers, proteins, and fatty acids [46].

In this respect, consumers generally require to be offered "all-natural" foods, and in order to provide them with this type of products, "natural ingredients" have been manufactured by biotechnological, membrane technology, and supercritical fluid extraction [24, 25].

2.3 Probiotics and prebiotics

Probiotics and prebiotics are part of the so-called foods that promote health. They are considered to be major nutrients that influence the physiology and gastrointestinal function [47].

According to the FAO/WHO definition, probiotics are defined as "live microorganisms which when administered in adequate amounts confer a health benefit on the host." Probiotics have beneficial effects on health through mechanisms of action, such as preventing colonization or pathogen adhesion, metabolism production, and immune system modulation by producing antibodies to immunoglobulins [48].

Probiotics, according to Gipson, are "a substrate that is selectively used by host microorganisms that confer a health benefit" [50]. The most used probiotic compounds are inulin, fructo-oligosaccharides, fructans, lactulose, and galacto-oligosaccharides [51].

3. Thermal technologies

3.1 Radio frequency heating

Radio frequency heating (dielectric heating) is a process by which a high-frequency radio wave is created using a generator for heating a dielectric material. Heat is quickly generated in the center of the food [52].

Each food product has specific dielectric properties, and these depend on the viscosity, temperature, chemical composition, and other physiological properties of the food [53]. It is applied in the food industry for continuous and batch heat processes [54]; rapid defrosting of frozen fish, meat, and other processed materials or foodstuff [55]; baking, disinfecting, and sanitizing dry food products (seeds, cereals, dried fruits, legumes); and sterilizing solid or viscous packaged foods [53, 54].

3.2 Microwave heating

Microwave heating is a thermal process that is performed with electromagnetic microwave radiation (1–100 GHz) and heat transfer. Any food that is exposed to the microwave is heated due to the electric and magnetic fields that generate heat [56].

The dielectric and magnetic properties of foods influence microwave activity. Due to the presence of water, microwave-treated liquid foods absorb electromagnetic energy very quickly [57].

Foods heat faster outside than indoors. Microwave heating is not uniform and results in nutrient losses due to the high temperature of the heated surface [58]. Microwave heating is used at home and at industrial level. At the household level, it is used for heating and thawing food, because it is a very fast method. In the food industry it is used for food drying, baking of bread and biscuits, pre-preparation and cooking of foods (cereals, meat, and meat products), defrosting food, bleaching vegetables, and pasteurizing and sterilizing fast food/liquid food, meals, and other food products [56, 59, 60].

3.3 Ohmic heating

Ohmic heating is a process of Joule heating, electro-heating, or electro-conductive heating. Electricity flows through the food. This process compared to microwave and radio frequency heating has no penetration depth limitation. The heating is carried out evenly, without damaging the nutrients in the food [61, 62].

Ohmic heating's applied products are for liquid foods containing large particles: slices of fruit in syrups and sauces, stews, soups, and heat-sensitive liquids [61, 63]. Food products subjected to ohmic heating treatment are high-quality, value-added, and shelf-stable products. This process is also suitable for defrosting, pasteurization, sterilization, extraction, dehydration, fermentation, evaporation, peeling, bleaching, packaging, and heating of food at serving temperature [62, 64, 65].

3.4 Infrared heating

The infrared process is between the ultraviolet energy region and the microwave. Infrared radiation is normally classified based on its spectral spectrum, in near-infrared (700–1400 nm), mid-infrared (1400–3000 nm), and far-infrared (3000–10,000 nm) regions [52, 66].

The far infrared is considered to be the most useful region for food processing, as most food components are absorbed by radiation in this region [66, 67]. Infrared heating is an indirect way of heat. At this stage electromagnetic energy enters food, adsorbs to the surface, and then converts to heat [58, 67].

Heating by radiant energy depends on the characteristics and color of the food. IR radiation is used to change the quality of food by changing the flavor, aroma, and the color of the food surface [58, 67].

Infrared heating is used in the food industry for roasting, cooking, baking, dehydrating, drying, pasteurizing, peeling, bleaching, and food processing [66, 68, 69]. The process of heating by IR is used in the food industry successfully and for 32 inactivation of lipoxygenase, lipases, α amylases, and enzymes responsible for the development of aromas and for the damage of fruits and vegetable and inactivation of bacteria, spores, yeast, and mold in both liquid and solid foods [70, 71].

Infrared heating can be combined for processes such as dehydration, cooking, baking, and freeze-drying [69, 72].

4. Nonthermal technologies

The term "nonthermal processing" is very often used to describe efficient processes at ambient or sublethal temperatures [73].

Developing and optimizing the processes of novel food preservation have become a major trend in food processing nowadays. Thus, those used to obtain minimally processed foods stand out alongside the ones based on emerging physical techniques (high-pressure processing, pulsed electric field processing, cold plasma treatment, ultrasound processing, irradiation, UV and pulsed light). These processes enable engineers and scientists to produce more nutritive, fresher, less processed, and safer foods [1, 7].

Minimally processed products gained more attention due to health considerations, in the last decade. In comparison to thermal processing, nonthermal techniques are beneficial to maintain the freshness, nutritional properties, flavor, and color attributes, especially some thermal unstable compounds such as ascorbic acid and polyphenols. More than that, novel nonthermal processing techniques show a potential application in the reduction of food immunoreactivity [74].

This processing method involves a loss in food quality, though thermal preservation methods provide safer foods. Hence, minimizing the degradation of food quality by limiting the damage that heat can produce on the food constitutes the main objective of the nonthermal preservation methods. They also imply the inactivation of those microorganisms and enzymes that are responsible for food degradation [7, 73].

4.1 High-pressure processing

High hydrostatic pressure was first used in Japan in the 1990s. It has been improved over the years and is now used worldwide. It's basically a cold pasteurization, which is traditionally used in the destruction and inactivation of spoilage and pathogenic microorganisms while preserving the quality of food. This technology is an energy-efficient and rapidly acting technological aid used in today's food processing [75–78]. The aim of this process is to extract a considerable amount of bioactive compounds from foods and the enhancement of their bioavailability [79, 80]. Sensory, nutritional, and functional properties of fruit and vegetable beverages have been slightly modified by means of the nonthermal processes in whose development and design scientists showed a great interest, in recent years [81]. As compared to conventional methods, processing has become cheaper in some cases as it requires a lower energy input, although the HHP equipment involves considerable costs [82].

High hydrostatic pressure processing (100–1000 MPa) is a minimal thermal technology applied to food products, often at room temperature [83]. This technology has been primarily focused as a substitute technology for heat processing. Heat processing is used to destroy or inhibit the activity of damaged microorganisms or enzymes. It can be successfully used at room temperature, reducing the thermal energy required for heat processing, to make food products without unwanted changes, such as nutritional and sensory characteristics. As a minimal thermal process, high hydrostatic pressure can be applied to heat-sensitive foods to guarantee microbiological safety and quality characteristics in minimally processed products. Therefore, the application of this process can provide high-quality pasteurized food products. Thus, high-pressure processing may provide fruit and vegetable products with suitable shelf life that maintain characteristics similar to fresh products, which consumers demand [81, 84, 85].

The critical factors in HHP process are shown in **Table 1**, and their magnitude will depend on the microorganism or enzyme subject to inactivation [86] (**Table 2**).

High hydrostatic pressure process preserves the freshness and taste of the product at a higher level, has low processing losses, and thus has a high product yield [87].

This technology has been analyzed, with good results, on a number of foods: beverages, juices, vegetables, fruits, meat products, fish and seafood, and ready-to-eat foods [88–90]. The technology is successfully used to preserve meat products [88] or to increase the shelf life of goat cheese and yogurt and reduces the allergenicity of milk and decreases cheese ripening time [91–93].

Conclusion. Advantages, limitations, and commercial applications of high-pressure processing are presented in **Table 3**.

Lactobacillus	***Bifidobacterium***	**Others**
L. acidophilus NCFM	*B. adolescentis*	*Enterococcus faecium*
L. bulgaricus	*B. animalis*	*Pediococcus pentosaceus*
L. casei	*B. breve*	*Saccharomyces boulardii*
L. delbrueckii	*B. bifidum*	
L. kefiranofaciens M1,	*B. lactis*	
L. paraplantarum	*B. longum*	
L. paracasei	*B. pseudocatenulatum*	
L. plantarum		
L. reuteri		
L. rhamnosus		
L. salivarius		

Table 1.
The most frequently used probiotics [49].

Pressure
Time required to achieve the treatment pressure
Time at an specific pressure
Treatment temperature
Final decompression time
Vessel temperature distribution at pressure
Initial product temperature
Product composition
Product pH
Packaging material integrity
Product water activity

Table 2.
Factors influencing the high hydrostatic pressure process [7, 86].

4.2 Pulsed electric field processing

This technique uses high-intensity pulsed electric field and involves pulses of high voltage (typically 20–80 kV/cm), for short periods of time (less than 1 second), which pass through the product placed between a set of electrodes inside a chamber, which are applied to fluid foods [94–96].

This main microbial process is used in order to inactivate some enzymes. It is also used to reduce the heating time of foods, and thus the degradation in the sensory and physical properties of foods is minimized [7]. In **Table 4**, a description of the different factors that affect the microbial inactivation with pulsed electric field processing can be found.

Pulsed electric field has been used in several processes: food processing, bioprocessing, inactivation of microorganisms, or permeabilization of food cells without thermal effects. The application of the processes used has good results for liquid and semisolid food products. Satisfactory results were obtained for the processing and preservation of foods such as milk, fruit juices, cooked meat, liquid eggs, and soups [94–96, 98].

Advantages	Limitations	Commercial applications
• No evidence of toxicity	• Little effect on food enzyme activity	• Kills vegetative bacteria (and spores at higher temperatures)
• Colors, flavors, and nutrients are preserved	• Some microbes may survive	• Pasteurization and sterilization of fruits, vegetables, meats, sauces, pickles, yoghurts, and salad dressings
• Reduced processing times	• Expensive equipment	• Potential for reduction or elimination of chemical preservatives
• Uniformity of treatment throughout food	• Foods should have approx. 40% free water for antimicrobial effect	• Decontamination of high-risk or high-value heat-sensitive ingredients
• Desirable texture changes possible	• Limited packaging options	
• In-package processing possible	• Regulatory issues to be resolved	

Table 3.
Advantages, limitations, and commercial applications of high-pressure processing [1].

Pulse width
Electric field intensity
Treatment temperature
Treatment time
Type of microorganism
Pulse wave shapes
pH
Concentration and growth stage of microorganism
Medium conductivity
Presence or absence of antimicrobials and ionic compounds
Medium ionic strength

Table 4.
Factors affecting the microbial inactivation with pulsed electric fields [7, 97].

However, pulsed electric field processing does not perform well for solid food products without air bubbles that have very low electrical conductivity [99].

In food processing it is used successfully for several types of fruit. The analyses have shown that they cause a minimal negative effect on the physical and sensory properties but increase the shelf life and the functional and textural properties of the juices [95, 100].

The process is successfully used for the production of French fries to reduce the cutting force required. In this situation, it has been shown that it inactivates microorganisms, preserving the nutritional and sensory quality of foods [96, 99, 101].

Conclusion. Advantages, limitations, and commercial applications of pulsed electric field processing are presented in **Table 5**.

4.3 Cold plasma treatment

Cold plasma is considered a modern technology, with nonthermal activity, which has been used in food processing in microbial inactivation and in the

Advantages	Limitations	Commercial applications
• There are no records of toxicity	• Unidentified activity on spores and enzymes	• For liquid foods
• Flavors, colors, and nutrients are preserved	• Hard to use with conductive materials	• Pasteurization of fruit juices, soups, liquid egg, and milk
• The treatment time is relatively short	• Satisfactory activity only for liquids or liquid particles	• Accelerated thawing
	• Satisfactory results only in combination with heat	• Decontamination of heat-sensitive foods
	• Foods may be affected by process of electrolysis	• Inactivates vegetative cells
	• Safety worry in local processing environment	
	• Energy activity not yet certain	
	• Regulatory issues are not clarified	
	• Unsatisfactory results in the presence of bubbles	
	• Safety and operational issues	

Table 5.
Advantages, limitations, and commercial applications of pulsed electric field [1].

decontamination of products from the food industry. This technology has been used because the energy consumption is lower and the demands on temperature are lower when compared to conventional processing methods [102, 103].

The cold plasma processing system is supported by a ceramic electrode and a high-frequency plasma generator [104]. Plasma is, in fact, a partially ionized gas consisting of reactive species, such as ions, electrons, UV photons, molecules, free radicals, and excited atoms. Plasma components can interact with proteins and modify their conformations [102, 105, 106].

In the food industry, cold plasma is used as a powerful disinfection tool for decontamination in packaging and after packaging of food products. It is also used for dry disinfection of solid and liquid foods (meat, fish, dry milk, sprouted seeds, herbs, spices, grains, and fresh products) [107–109].

Cold plasma treatment is a fast technology and does not leave toxic residues or post-processing exhaust gases. Even if it has certain advantages, we must take into account the aspects regarding the nutritional content, the color, the texture, the chemical changes, and the general quality of the food. These aspects are influenced, cold plasma treatments are not yet widely used, and more studies are needed [103, 108, 110].

Conclusion. Advantages, limitations, and commercial applications of cold plasma treatment are presented in **Table 6**.

4.4 Ultrasound

Sound waves whose frequency exceeds the hearing limit of the human ear (~20 kHz) are known as ultrasound. Ultrasound is one of the recently developed technologies in view of maximizing quality, minimizing processing, and ensuring the safety of food products. Ultrasound is applied in food processing in order to improve processes such as food preservation, mass transfer, manipulation of texture, assistance of thermal treatments, and food analysis [111].

Advantages	Limitations	Commercial applications
• Effective with temperature sensitive products	• There is no commercial tool available for disinfecting and sanitizing food products and packaging materials	• Disinfection of processing equipment. Shelf life extension
• Reduce cross-contamination and the establishment of biofilms on equipment	• Applied by various research organizations and universities, but not by industry	• Food processing equipment, food packaging, food contact surfaces, preservation
• Minimal effects on food quality and appearance of the product	• Interaction of electronically excited molecules with the food or packaging materials needs to be identified	• Inactivates surface spores and microflora on packaging food/ materials surfaces
• No shadowing effect ensuring all parts of a product are treated	• Modification of food packaging polymers is expected	• Technology of decontamination for mild surface such as cut fresh meat and vegetables
	• No potential scale up to pilot plant level for food industry yet	• Irregularly shaped packages such as bottles can be effectively treated, contrary to technologies such as UV or pulsed light where shadowing occurs
	• Stability for large-scale commercial operations is not clear	
	• Spore inactivation mechanism is unknown	
	• Regulatory issues	

Table 6.
Advantages, limitations, and commercial applications of cold plasma treatment [1].

In food processing, ultrasound is applied for analysis and quality control, and given the frequency range, it can be divided into low and high energy.

Ultrasonic processing is successfully used for food processing and preservation processes. Also, ultrasound is used for crystallization, drying, emulsification, solubility, homogenization, dispersion, improving texture, or modifying the viscosity and fermentation process [112–116].

It is successfully used to increase microbial safety in fruit juices [117, 118]. The use of ultrasound for the extraction of bioactive compounds from seeds, plants, or food has shown an increase in extraction efficiency [113, 114].

Although, this technology has advantages in food preservation and extraction of some compounds, studies have shown that it diminishes the quality of food: nutritional value, aroma, and color [119, 120]. Consequently, it is necessary to deepen studies for large-scale use in food processing.

Conclusion. Advantages, limitations, and commercial applications of ultrasound processing are presented in **Table 7**.

4.5 Irradiation

Radiation is a process used to conserve food. This is a nonthermal process that lowers or eliminates microorganisms and does not destroy the properties of food. This process is considered by 55 countries to be safe [121, 122] if 1 of the 3 approved irradiation processes is used: gamma rays, X-rays, or electron beams [123].

Advantages	Limitations	Commercial applications
• Increased heat transfer	• Depth of penetration affected by solids and air in the product	• Effective against vegetative cells, spores, and enzymes
• Little adaptation required of existing processing plant	• Complex mode of action	• Effective tool for microbial inactivation
• Reduction of process times and temperatures	• Needs to be used in combination with another process (e.g., heating)	• Minimal effect on the ascorbic acid content during processing
• Can be used alone or in combination with heat and/or pressure	• Potential problems with scaling-up plant	• Enhances extraction yield
• Batch or continuous operation	• Possible damage by free radicals	• Fruit juices preservation
• Higher throughput and lower energy consumption	• Unwanted modification of food structure and texture	
• Achieves a desired 5 log for foodborne pathogens in fruit juices	• Negatively modify some food properties including flavor, color, or nutritional value	
	• Possible modification of food structure and texture	

Table 7.
Advantages, limitations, and commercial applications of ultrasound processing [1].

Advantages	Limitations	Commercial applications
• Reliable and energy efficient	• Localized risks from radiation	• Suitable for nonmicrobial applications (e.g., sprout inhibition)
• Excellent penetration into foods	• Poor consumer understanding	• Insecticidal
• Improvement in flavor in some foods	• High capital cost	• Suitable for sterilization
• Suitable for large-scale production	• Difficult to detect	• Packaging
• Little loss of food quality	• Higher doses may produce radiation-induced degradation products	• Appropriate for fruits, vegetables, herbs, spices, meat and fish preservation
• Negligible or subtle losses of bioactive compounds	• Changes in flavor due to oxidation	• Suitable for raw, dry foods or processed food
• Minimal modification in the flavor, color, nutrients, taste, and other quality attributes of food	• Formation of free radicals	
• No increase in food temperature during processing		

Table 8.
Advantages, limitations, and commercial applications of irradiation [1].

Gamma or X-rays rapidly enter food and inactivate microorganisms. They are high-frequency waves; they do not generate heat and thus the quality of the food remains intact [124–126].

Following irradiation, food undergoes minimal changes in the content of nutrients, aroma, color, and taste, considered insignificant [122, 127].

In the food industry, irradiation is used to reduce post-harvest losses, to preserve the color of meat, and to inhibit germ formation in products such as potatoes. It is also used to control post-packaging contamination for several types of foods (vegetables, fruits, spices, cereals, fish) [124, 128].

According to studies, not all foods are suitable for irradiation, for example, milk and foods that are high in lipids and vitamins [129]. There are contradictory studies on the influence of irradiation of food products and packaging; therefore it is a topical topic for new studies [125].

Conclusion. Advantages, limitations, and commercial applications of irradiation are presented in **Table 8.**

4.6 UV and pulsed light

Intense and short-duration pulses of broad spectrum "white light" (UV in the near-infrared region) are used in the method of pulsed and UV light, which has

Advantages	Limitations	Commercial applications
• No thermal effect, so quality and nutrient content are retained	• Pulsed light—mostly suitable for liquid foods and surface of solid foods, hence limiting its application	• Shelf life extension of ready-to-eat cooked meat products
• Can be applied with other nonthermal processing technologies	• Pulsed light—packaging materials for irradiation should be chemically stable	• Alternative treatment to thermal pasteurization of fresh juices
• Maintains food texture and nutrients	• Pulsed light—the material should be transparent in order to allow the light to pass into the food	• Bacterial inactivation in fruit juices and milk
• Unlike chemical biocides, UV does not alter the chemical composition, taste, odor, or pH of the product and leave no toxins or residues into the process	• Pulsed light—the mechanism by which pulsed light induces cell death is yet to be fully explained	• Surface decontamination of eggs and chicken
• Neither increases the temperature of the product nor produces undesirable organoleptic changes	• UV—dose response behavior of food pathogens in viscous liquid foods needs to be developed	• Decontamination of food powders
	• UV—more kinetic inactivation data for pathogen and spoilage microorganisms is required to predict UV disinfection rates on food surfaces	• Decontamination of food processing equipment
		• Decontamination of air and surfaces
		• Water sterilization and wastewater disinfection
		• Mitigation of allergen from food

Table 9.
Advantages, limitations, and commercial applications of UV and pulsed light [1].

been made use of lately in food preservation, For most applications, a high level of microbial inactivation is provided within a fraction of a second, just by applying some flashes onto the food products. Ultraviolet light and pulsed light are modern techniques used to increase food safety. These techniques are minimally invasive, maintain the nutritional qualities and sensory appearance of foods, and at the same time extend the shelf life [130–132]. UV technology uses shorter wavelength (100–380 nm) while pulsed light operates on a broad spectrum of light (180–1100 nm) [133].

Because of its poor penetration level, this technology has been used in sterilization and the reduction of microbial load on food processing equipment, food surfaces, or packaging materials. However, this method with UV light is also used in the pasteurization process of fruit juices [134].

The application of UV technology has been used for the first time in Europe to disinfect municipal drinking water as an alternative to chloride. It is now used globally for the treatment of drinking water, processing water, wastewater, and industrial water [132, 135]. This treatment is used as an alternative method for the thermal pasteurization of fresh juices. It is also used as a disinfection method in the food industry. Research has shown that it is effective against bacterial pathogens, does not increase the temperature of the treated products, and does not produce unwanted organoleptic changes [136, 137].

Pulsed light is a nonthermal technology and is probably the best alternative method for decontaminating surfaces in the food industry and food packaging. This technology has the role of sterilizing or inactivating microorganisms on work surfaces, equipment, packaging materials, and equipment [131]. According to studies, pulsed light inactivates bacteria, viruses, and fungi even faster and more effectively than continuous UV treatment [130, 133].

Conclusion. Advantages, limitations, and commercial applications of UV and pulsed light are presented in **Table 9**.

5. Conclusion

With the new trends of the modern world, food consumers come with new demands on food. They consider that a food is not enough to satisfy the need for hunger, but it must be as nutritious as possible and can prevent certain diseases. The processing methods used in the food industry, presented in this chapter, are the newest methods available. They try to be minimally invasive, to ensure food safety, and to maintain their nutritional, sensory, and structural qualities. The technologies presented are used in food processing, but they can be improved and require future studies to deepen the existing information.

Acknowledgements

Special thanks are addressed to Ms. Katarina Pausic and Ms. Marijana Francetic, who assisted in the arrangement of the book and scheduling of our activities.

Conflict of interest

Author declares there is no conflict of interest.

Author details

Romina Alina Marc
Faculty of Food Science and Technology, University of Agricultural Sciences and Veterinary Medicine Cluj-Napoca, Romania

*Address all correspondence to: romina.vlaic@usamvcluj.ro

References

[1] Priyadarshini A, Rajauria G, O'Donnell CP, Tiwari BK. Emerging food processing technologies and factors impacting their industrial adoption. Critical Reviews in Food Science and Nutrition. 2019;**59**(19):3082-3101. DOI: 10.1080/10408398.2018.1483890

[2] Knoerzer K, Juliano P, Roupas P, Versteeg C. Innovative Food Processing Technologies: Advances in Multiphysics Simulation. Chichester, West Sussex, UK: John Wiley & Sons and Institute of Food Technologists; 2011. DOI: 10.1002/9780470959435

[3] Boekel V, Martinus VF, Pellegrini N, Stanton C, Scholz G, Lalljie S, et al. A review on the beneficial aspects of food processing. Molecular Nutrition & Food Research. 2010;**54**(9):1215-1247

[4] Falguera V, Aliguer N, Falguera M. An integrated approach to current trends in food consumption: Moving toward functional and organic products? Food Control. 2012;**26**(2):274-281. DOI: 10.1016/j.foodcont.2012.01.051

[5] Moskowitz HR, Beckley JH, Resurreccion AVA. Sensory and Consumer Research in Food Product Design and Development. Ames, Iowa, USA: John Wiley & Sons; 2012. DOI: 10.1002/9781119945970

[6] Moubarac J-C, Parra DC, Cannon G, Monteiro CA. Food classification systems based on food processing: Significance and implications for policies and actions: A systematic literature review and assessment. Current Obesity Reports. 2014;**3**(2): 256-272. DOI: 10.1007/s13679-014-0092-0

[7] Señorans FJ, Ibáñez E, Cifuentes A. New trends in food processing. Critical Reviews in Food Science and Nutrition. 2003;**43**(5):507-526. DOI: 10.1080/10408690390246341

[8] Soon JM, Brazier AKM, Wallace CA. Determining common contributory factors in food safety incidents—A review of global outbreaks and recalls 2008–2018. Trends in Food Science & Technology. 2020;**97**:76-87. DOI: 10.1016/j.tifs.2019.12.030

[9] Nyarugwe SP, Linnemann AR, Luning PA. Prevailing food safety culture in companies operating in a transition economy—Does product riskiness matter? Food Control. 2019; **107**:106803. DOI: 10.1016/j.foodcont.2019.106803

[10] Wenzl T, Simon R, Anklam E, Kleiner J. Analytical methods for polycyclic aromatic hydrocarbons (PAHs) in food and the environment needed for new food legislation in the European Union. TrAC Trends in Analytical Chemistry. 2006;**25**(7): 716-725. DOI: 10.1016/j.trac.2006.05.010

[11] Eriksson D, Custers R, Björnberg KE, Hansson SO, Purnhagen K, Qaim M, et al. Visser options to reform the european union legislation on GMOs: Scope and definitions. Trends in Biotechnology. 2020;**38**(3):231-234

[12] Ollinger ME, Muth MK, Karns S, Choice Z. Food safety audits, plant characteristics, and food safety technology use in meat and poultry plants. USDA-ERS Economic Information Bulletin. 2011;**82**:1-44. DOI: 10.2139/ssrn.2131423

[13] Yeung R, Morris J. Food safety risk: Consumer perception and purchase behaviour. British Food Journal. 2001; **103**(3):170-187. DOI: 10.1108/00070700110386728

[14] Menrad K. Innovations in the food industry in Germany. Research Policy.

2004;**33**(6-7):845-878. DOI: 10.1016/j.respol.2004.01.012

[15] Samoggia A, Riedel B. Assessment of nutrition-focused mobile apps' influence on consumers' healthy food behaviour and nutrition knowledge. Food Research International. 2019; **108766**:1-30. DOI: 10.1016/j.foodres.2019.108766

[16] Black MM, Delichatsios HK, Story MT, editors. Nutrition Education: Strategies for Improving Nutrition and Healthy Eating in Individuals and Communities. Switzerland/S. Karger AG., Basel: Nestlé Nutr Inst Workshop Ser. Nestlé Nutrition Institute; 2020. p. 92. DOI: 10.1159/000499552

[17] Bowman BA, Mokdad AH. Addressing nutrition and chronic disease: Past, present, and future research directions. Food and Nutrition Bulletin. 2020;**41**(1):3-7. DOI: 10.1177/0379572119893904

[18] Annunziata A, Vecchio R. Functional foods development in the European market: A consumer perspective. Journal of Functional Foods. 2011;**3**(3):223-228. DOI: 10.1016/j.jff.2011.03.011

[19] Granato D, Barba FJ, Kovačević DB, Lorenzo JM, Cruz AG, Putnik P. Functional foods: Product development, technological trends, efficacy testing, and safety. Annual Review of Food Science and Technology. 2020;**11**(1): 93-118. DOI: 10.1146/annurev-food-032519-051708

[20] Hasler CM. Functional foods: Their role in disease prevention and health promotion. Food Technology. 1998;**52**: 63-70

[21] Sloan AE. The top ten functional food trends. Food Technology. 2000;**54**: 33-62

[22] Sloan AE. The top ten functional food trends. Food Technology (Chicago). 2014;**68**(4):22-45

[23] Siró I, Kápolna E, Kápolna B, Lugasi A. Functional food. Product development, marketing and consumer acceptance—A review. Appetite. 2008; **51**(3):456-467. DOI: 10.1016/j.appet.2008.05.060

[24] Bagchi D. Nutraceutical and Functional Food Regulations in the United States and around the World. 3rd ed: USA: Academic Press Elsevier; 2019. DOI: 10.1016/B978-0-12-373901-8.X0001-7

[25] Pszczola DE. The ABCs of nutraceutical ingredients. Food Technology. 1998;**52**:30-37

[26] Gonzalez Candelas L, Gil JV, Lamuela Raventos RM, Ramon D. The use of transgenic yeasts expressing a gene encoding a glycosyl hydrolase as a tool to increase resveratrol content in wine. International Journal of Food Microbiology. 2000;**59**:179-183

[27] Nakamura R, Matsuda T. Rice allergenic protein and molecular-genetic approach for hypoallergenic rice. Bioscience, Biotechnology, and Biochemistry. 1996;**60**:1215-1221

[28] Szczodrak J. Hydrolysis of lactose in whey permeate by immobilized beta-galactosidase from *Penicillium notatum*. Acta Biotechnologica. 1999;**19**:235-250

[29] Ki Teak L, Foglia TA. Synthesis, purification, and characterization of structured lipids produced from chicken fat. Journal of the American Oil Chemists' Society. 2000;**77**:1027-1034

[30] Girard B, Fukumoto LR. Membrane processing of fruit juices and beverages: A review. Critical Reviews in Food Science and Nutrition. 2000;**40**(2): 91-157. DOI: 10.1080/10408690091189293

[31] Marella C. Application of membrane separation technology for developing novel dairy food ingredients. Journal of Food Processing & Technology. 2013;**04**(09):1-5. DOI: 10.4172/2157-7110.1000269

[32] Mohammad AW, Ng CY, Lim YP, Ng GH. Ultrafiltration in food processing industry: Review on application, membrane fouling, and fouling control. Food and Bioprocess Technology. 2012;**5**(4):1143-1156. DOI: 10.1007/s11947-012-0806-9

[33] Salehi F. Current and future applications for nanofiltration technology in the food processing. Food and Bioproducts Processing. 2014;**92**(2): 161-177. DOI: 10.1016/j.fbp.2013.09.005

[34] Mondor M, Ippersiel D, Lamarche F. Electrodialysis in food processing. In: Boye J, Arcand Y, editors. Green Technologies in Food Production and Processing. Food Engineering Series. Boston, MA: Springer; 2012. ISBN: 978-1-4614-1586-2

[35] Wang B, Timilsena YP, Blanch E, Adhikari B. Lactoferrin: Structure, function, denaturation and digestion. Critical Reviews in Food Science and Nutrition. 2019;**59**(4):580-596. DOI: 10.1080/10408398.2017.1381583

[36] Sharif KM, Rahman MM, Azmir J, Mohamed A, Jahurul MHA, Sahena F, et al. Experimental design of supercritical fluid extraction—A review. Journal of Food Engineering. 2014;**124**: 105-116. DOI: 10.1016/j.jfoodeng.2013.10.003

[37] Katz SN, Spence JE, O'Brien MJ, Skiff RH, Vogel GJ, Prasad R. Decaffeination of coffee. Europ. Patent Appl. EP 0 424 579 A1, EP 89-310933. 1991

[38] Zabot GL. Chapter 11—Decaffeination using supercritical carbon dioxide. Green Sustainable Process for Chemical and Environmental Engineering and Science. Amsterdam, Netherland: Elsevier Inc.; 2020. DOI: 10.1016/b978-0-12-817388-6.00011-8

[39] Medina I, Martinez JL. Dealcoholisation of cider by supercritical extraction with carbon dioxide. Journal of Chemical Technology and Biotechnology. 1997;**68**:14-18

[40] Gamse T, Rogler I, Marr R. Supercritical CO_2 extraction for utilisation of excess wine of poor quality. The Journal of Supercritical Fluids. 1999;**14**(2):123-128. DOI: 10.1016/s0896-8446(98)00114-4

[41] Rizvi SSH, editor. Super Critical Fluid Processing of Food and Biomaterials. Glasgow: Blackie Academic & Professional; 1994

[42] Pavlić B, Pezo L, Marić B, Tukuljac LP, Zeković Z, Solarov MB, et al. Supercritical fluid extraction of raspberry seed oil: Experiments and modelling. The Journal of Supercritical Fluids. 2019;**157**:104687. DOI: 10.1016/j.supflu.2019.104687

[43] Lee W-J, Suleiman N, Hadzir NHN, Chong G-H. Chapter 12—Supercritical fluids for the extraction of oleoresins and plant phenolics. Green Sustainable Process for Chemical and Environmental Engineering and Science. Amsterdam, Netherland: Elsevier Inc.; 2020. DOI: 10.1016/b978-0-12-817388-6.00012-x

[44] Gallego R, Bueno M, Herrero M. Sub- and supercritical fluid extraction of bioactive compounds from plants, food-by-products, seaweeds and microalgae—An update. TrAC Trends in Analytical Chemistry. 2019;**116**:198-213. DOI: 10.1016/j.trac.2019.04.030

[45] Liu L, Zhang Y, Zhou Y, Li G, Yang G, Feng X. The application of

supercritical fluid chromatography in Food quality and Food safety: An overview. Critical Reviews in Analytical Chemistry. 2019;**50**:2,136-160. DOI: 10.1080/10408347.2019.1586520

[46] Vlaic RA, Mureşan CC, Muste S, Mureşan A, Muresan V, Suharoschi R, et al. Food Fortification through Innovative Technologies. London, UK: Editura IntechOpen; 2019. DOI: 10.5772/intechopen.82249

[47] Salminen S. Probiotics: Scientific support for use. Food Technology. 1999; **53**:66

[48] Chugh B, Kamal-Eldin A. Bioactive compounds produced by probiotics in food products. Current Opinion in Food Science. 2020;1-5. DOI: 10.1016/j. cofs.2020.02.003

[49] Terpou A, Papadaki A, Lappa IK, Kachrimanidou V, Bosnea LA, Kopsahelis N. Probiotics in food systems: Significance and emerging strategies towards improved viability and delivery of enhanced beneficial value. Nutrients. 2019;**11**(7):1591. DOI: 10.3390/nu11071591

[50] Gibson GR, Hutkins R, Sanders ME, Prescott SL, Reimer RA, Salminen SJ, et al. Expert consensus document: The International Scientific Association for Probiotics and Prebiotics (ISAPP) consensus statement on the definition and scope of prebiotics. Nature reviews. Gastroenterology & Hepatology. 2017; **14**:491-502

[51] Markowiak P, Śliżewska K. Effects of probiotics, prebiotics, and synbiotics on human health. Nutrients. 2017;**9**: 1021

[52] Maloney N, Harrison M. Chapter 8 —Advanced heating technologies for food processing A2. In: Leadley CE, editor. Innovation and Future Trends in Food Manufacturing and Supply Chain Technologies. Chester, UK: Woodhead Publishing Series in Food Science, Technology and Nutrition; 2016. pp. 203-256. eBook ISBN: 9781782424703

[53] Uyar R, Bedane TF, Erdogdu F, Palazoglu TK, Farag KW, Marra F. Radio-frequency thawing of food products—A computational study. Journal of Food Engineering. 2015;**146**: 163-171. DOI: 10.1016/j.jfoodeng.2014. 08.018

[54] Huang Z, Marra F, Wang S. A novel strategy for improving radio frequency heating uniformity of dry food products using computational modeling. Innovative Food Science & Emerging Technologies. 2016;**34**:100-111. DOI: 10.1016/j.ifset.2016.01.005

[55] Ha J-W, Kim S-Y, Ryu S-R, Kang D-H. Inactivation of Salmonella enterica serovar Typhimurium and Escherichia coli O157:H7 in peanut butter cracker sandwiches by radio-frequency heating. Food Microbiology. 2013;**34**(1):145-150. DOI: 10.1016/j. fm.2012.11.018

[56] Leonelli C, Mason TJ. Microwave and ultrasonic processing: Now a realistic option for industry. Chemical Engineering and Processing: Process Intensification. 2010;**49**(9):885-900. DOI: 10.1016/j.cep.2010.05.006

[57] Ahmed J, Ramaswamy HS. Microwave pasteurization and sterilization of foods. In: Food Science and Technology. Vol. 167. New York: Marcel Dekker; 2004. p. 691

[58] Kodandaram Reddy D, Waghray K, Sathyanarayana SV. Infrared Heating - A New Green Technology for Process Intensification in Drying of Purslane Leaves to Reduce the Thermal Losses. Springer Nature Switzerland, 2020. DOI: 10.1007/978-3-030-24314-2_56

[59] Liu W, Lanier TC. Rapid (microwave) heating rate effects on texture, fat/water holding, and

microstructure of cooked comminuted meat batters. Food Research International. 2016;**81**:108-113. DOI: 10.1016/j.foodres.2016.01.005

[60] Monteiro RL, Carciofi BAM, Laurindo JB. A microwave multi-flash drying process for producing crispy bananas. Journal of Food Engineering. 2016;**178**:1-11. DOI: 10.1016/j.jfoodeng.2015.12.024

[61] Wongsa-Ngasri P, Sastry SK. Tomato peeling by ohmic heating: Effects of lye-salt combinations and post-treatments on weight loss, peeling quality and firmness. Innovative Food Science & Emerging Technologies. 2016;**34**:148-153. DOI: 10.1016/j.ifset.2016.01.013

[62] Varghese K, Shiby M, Pandey C, Radhakrishna K, Bawa AS. Technology, applications and modelling of ohmic heating: A review. Journal of Food Science and Technology. 2014;**51**(10): 2304-2317. DOI: 10.1007/s13197-012-0710-3

[63] Saxena J, Makroo HA, Srivastava B. Optimization of time electric field combination for PPO inactivation in sugarcane juice by ohmic heating and its shelf life assessment. LWT—Food Science and Technology. 2016;**71**: 329-338. DOI: 10.1016/j.lwt.2016.04.015

[64] Duygu B, Ümit G. Application of ohmic heating system in meat thawing. Procedia—Social and Behavioral Sciences. 2015;**195**:2822-2828. DOI: 10.1016/j.sbspro.2015.06.400

[65] Fowler MR, Park JW. Effect of salmon plasma protein on Pacific whiting surimi gelation under various ohmic heating conditions. LWT—Food Science and Technology. 2015; **61**(2):309-315. DOI: 10.1016/j.lwt.2014.12.049

[66] Rastogi NK. Infrared heating of foods and its combination with electron beam processing. Electron Beam Pasteurization and Complementary Food Processing Technologies. Chester, UK: Woodhead Publishing Series in Food Science, Technology and Nutrition; 2015:61-82. DOI: 10.1533/9781782421085.2.61

[67] Wang Y, Li X, Sun G, Dong L, Pan Z. A comparison of dynamic mechanical properties of processing-tomato peel as affected by hot lye and infrared radiation heating for peeling. Journal of Food Engineering. 2014;**126**: 27-34. Available from: https://doi.org/10.1016/j.jfoodeng.2013.10.032

[68] Moreirinha C, Almeida A, Saraiva JA, Delgadillo I. High-pressure processing effects on foodborne bacteria by mid-infrared spectroscopy analysis. LWT. 2016;**73**:212-218. DOI: 10.1016/j.lwt.2016.05.041

[69] Huang H-W, Hsu C-P, Yang BB, Wang C-Y. Potential utility of high-pressure processing to address the risk of food allergen concerns. Comprehensive Reviews in Food Science and Food Safety. 2014;**13**(1): 78-90

[70] Huang H-W, Hsu C-P, Yang BB, Wang C-Y. Potential utility of high-pressure processing to address the risk of food allergen concerns. Comprehensive Reviews in Food Science and Food Safety. 2013; **13**(1):78-90. DOI: 10.1111/1541-4337.12045

[71] Bermúdez-Aguirre D, Barbosa-Cánovas GV. An update on high hydrostatic pressure, from the laboratory to industrial applications. Food Engineering Reviews. 2010;**3**(1): 44-61. DOI: 10.1007/s12393-010-9030-4

[72] Mao W, Oshima Y, Yamanaka Y, Fukuoka M, Sakai N. Mathematical simulation of liquid food pasteurization using far infrared radiation heating equipment. Journal of Food

Engineering. 2011;**107**(1):127-133. DOI: 10.1016/j.jfoodeng.2011.05.024

[73] Pereira RN, Vicente AA. Environmental impact of novel thermal and non-thermal technologies in food processing. Food Research International. 2010;**43**(7):1936-1943. DOI: 10.1016/j.foodres.2009.09.013

[74] Rahaman T, Vasiljevic T, Ramchandran L. Effect of processing on conformational changes of food proteins related to allergenicity. Trends in Food Science & Technology. 2016;**49**:24-34. DOI: 10.1016/j.tifs.2016.01.001

[75] Pingen S, Sudhaus N, Becker A, Krischek C, Klein G. High pressure as an alternative processing step for ham production. Meat Science. 2016;**118**: 22-27. DOI: 10.1016/j.meatsci.2016.03.014

[76] Hribar J, Pozrl T, Vidrih R. Novel technologies in fruit and vegetable processing. Croatian Journal of Food Science and Technology. 2018;**10**(1): 112-117. DOI: 10.17508/CJFST.2018.10.1.14

[77] Mújica-Paz H, Valdez-Fragoso A, Samson CT, Welti-Chanes J, Torres JA. High-pressure processing technologies for the pasteurization and sterilization of foods. Food and Bioprocess Technology. 2011;**4**(6):969-985. DOI: 10.1007/s11947-011-0543-5

[78] Gómez-Maqueo A, Ortega-Hernández É, Serrano-Sandoval SN, et al. Addressing key features involved in bioactive extractability of vigor prickly pears submitted to high hydrostatic pressurization. Journal of Food Process Engineering. 2019;**43**(1): 1-11. DOI: 10.1111/jfpe.13202

[79] Barba FJ, Esteve MJ, Frigola A. Physicochemical and nutritional characteristics of blueberry juice after high pressure processing. Food Research International. 2013;**50**(2):545-549. DOI: 10.1016/j.foodres.2011.02.038

[80] Gómez-Maqueo A, García-Cayuela T, Welti-Chanes J, Cano MP. Enhancement of anti-inflammatory and antioxidant activities of prickly pear fruits by high hydrostatic pressure: A chemical and microstructural approach. Innovative Food Science & Emerging Technologies. 2019;**54**: 132-142. DOI: 10.1016/j.ifset.2019.04.002

[81] Bevilacqua A, Petruzzi L, Perricone M, Speranza B, Campaniello D, Sinigaglia M, et al. Nonthermal technologies for fruit and vegetable juices and beverages: Overview and advances. Comprehensive Reviews in Food Science and Food Safety. 2018;**17**(1): 2-62. DOI: 10.1111/1541-4337.12299

[82] Wang CY, Huang HW, Hsu CP, Yang BB. Recent advances in food processing using high hydrostatic pressure technology. Critical Reviews in Food Science and Nutrition. 2016;**56**(4): 527-540. DOI: 10.1080/10408398.2012.745479

[83] Guerrero-Beltrán JA, Barbosa-Cánovas GV, Swanson BG. High hydrostatic pressure processing of fruit and vegetable products. Food Reviews International. 2005;**21**(4):411-425. DOI: 10.1080/87559120500224827

[84] Knorr D. Effects of high-hydrostatic-pressure processes on food safety and quality. Food Technology. 1993;**47**:56-161

[85] Welti-Chanes J, San Martín-González F, Barbosa-Cánovas GV. Water and Biological Structures at High Pressure. Water Properties, Pharmaceutical, and Biological Materials. Boca Ratón, FL: CRC Press; 2006. pp. 205-231

[86] Smelt JPPM. Recent advances in the microbiology of high pressure processing. Trends in Food Science and Technology. 1998;**9**:152-158

[87] Tsevdou MS, Eleftheriou EG, Taoukis PS. Transglutaminase treatment of thermally and high pressure processed milk: Effects on the properties and storage stability of set yoghurt. Innovative Food Science & Emerging Technologies. 2013;**17**:144-152. DOI: 10.1016/j.ifset.2012.11.004

[88] Tribst AAL, de Castro Leite Júnior BR, de Oliveira MM, Cristianini M. High pressure processing of cocoyam, Peruvian carrot and sweet potato: Effect on oxidative enzymes and impact in the tuber color. Innovative Food Science & Emerging Technologies. 2016;**34**:302-309. DOI: 10.1016/j.ifset.2016.02.010

[89] Khan MA, Ali S, Abid M, Ahmad H, Zhang L, Tume RK, et al. Enhanced texture, yield and safety of a ready-to-eat salted duck meat product using a high pressure-heat process. Innovative Food Science & Emerging Technologies. 2014;**21**:50-57. DOI: 10.1016/j.ifset.2013.10.008

[90] Evert-Arriagada K, Hernández-Herrero MM, Guamis B, Trujillo AJ. Commercial application of high-pressure processing for increasing starter-free fresh cheese shelf-life. LWT—Food Science and Technology. 2014;**55**(2):498-505. DOI: 10.1016/j.lwt.2013.10.030

[91] Pingen S, Sudhaus N, Becker A, Krischek C, Klein G. High pressure as an alternative processing step for ham production. Meat Science. 2016;**118**:22-27. DOI: 10.1016/j.meatsci.2016.03.014

[92] Zhou Y, Karwe MV, Matthews KR. Differences in inactivation of *Escherichia coli* O157:H7 strains in ground beef following repeated high pressure processing treatments and cold storage. Food Microbiology. 2016;**58**:7-12. DOI: 10.1016/j.fm.2016.02.010

[93] Barba FJ, Parniakov O, Pereira SA, Wiktor A, Grimi N, Boussetta N, et al. Current applications and new opportunities for the use of pulsed electric fields in food science and industry. Food Research International. 2015;**77**(4):773-798. DOI: 10.1016/j.foodres.2015.09.015

[94] Toepfl S, Siemer C, Saldaña-Navarro G, Heinz V. Overview of pulsed electric fields processing for food. In: Sun D-W, editor. Emerging Technologies for Food Processing. 2nd ed. San Diego: Academic Press; 2014. pp. 93-114

[95] Mohamed MEA, Eissa AHA. Pulsed Electric Fields for Food Processing Technology: Rijeka INTECH Open Access Publisher. 2012; DOI: 10.5772/48678

[96] Ma Q, Hamid N, Oey I, Kantono K, Faridnia F, Yoo M, et al. Effect of chilled and freezing pre-treatments prior to pulsed electric field processing on volatile profile and sensory attributes of cooked lamb meats. Innovative Food Science & Emerging Technologies. 2016;**37**:359-374. DOI: 10.1016/j.ifset.2016.04.009

[97] Barbosa-Canovas G, Pothakamury UR, Palou E, Swanson BG, editors. Nonthermal Preservation of Foods. New York: Marcel Dekker, Inc.; 1998

[98] Agcam E, Akyildiz A, Akdemir Evrendilek G. A comparative assessment of long-term storage stability and quality attributes of orange juice in response to pulsed electric fields and heat treatments. Food and Bioproducts Processing. 2016;

99:90-98. DOI: 10.1016/j.fbp.2016.04.006

[99] Griffiths MW, Walkling-Ribeiro M. Chapter 7—Pulsed electric field processing of liquid foods and beverages. Emerging Technologies for Food Processing. 2nd ed. USA: Academic Press; 2014:115-145. DOI: 10.1016/b978-0-12-411479-1.00007-3

[100] Shakhova NB, Yurmazova TA, Hoang TT, Anh NT. Pulsed electric discharge in active metallic grains for water purification processes. Procedia Chemistry. 2015;**15**:292-300. DOI: 10.1016/j.proche.2015.10.047

[101] Niemira BA. Chapter 18—Decontamination of foods by cold plasma A2. In: Sun D-W, editor. Emerging Technologies for Food Processing. 2nd ed. San Diego: Academic Press; 2014. pp. 327-333. DOI: 10.1016/b978-0-12-411479-1.00018-8

[102] Ekezie FGC, Cheng JH, Sun DW. Effects of nonthermal food processing technologies on food allergens: A review of recent research advances. Trends in Food Science & Technology. 2018;**74**:12-25. DOI: 10.1016/j.tifs.2018.01.007

[103] Dong X, Wang J, Raghavan V. Critical reviews and recent advances of novel non-thermal processing techniques on the modification of food allergens. Critical Reviews in Food Science and Nutrition. 2020;**12**:1-15. DOI: 10.1080/10408398.2020.1722942

[104] Chizoba Ekezie F-G, Cheng J-H, Sun D-W. Effects of mild oxidative and structural modifications induced by argon plasma on physicochemical properties of actomyosin from king prawn (*Litopenaeus vannamei*). Journal of Agricultural and Food Chemistry. 2018;**66**(50):13285-13294. DOI: 10.1021/acs.jafc.8b05178

[105] Inagaki N. Plasma Surface Modification and Plasma Polymerization. Lancaster: CRC Press; 2014

[106] Tolouie H, Mohammadifar MA, Ghomi H, Hashemi M. Cold atmospheric plasma manipulation of proteins in food systems. Critical Reviews in Food Science and Nutrition. 2018;**58**(15):2583-2597. DOI: 10.1080/10408398.2017.1335689

[107] Jayasena DD, Kim HJ, Yong HI, Park S, Kim K, Choe W, et al. Flexible thin-layer dielectric barrier discharge plasma treatment of pork butt and beef loin: Effects on pathogen inactivation and meat quality attributes. Food Microbiology. 2015;**46**:51-57. DOI: 10.1016/j.fm.2014.07.009

[108] Korachi M, Ozen F, Aslan N, Vannini L. Maria Elisabetta Guerzoni, Davide Gottardi, and Fatma Yesim Ekinci, Biochemical changes to milk following treatment by a novel, cold atmospheric plasma system. International Dairy Journal. 2015;**42**:64-69. DOI: 10.1016/j.idairyj.2014.10.006

[109] Lee H, Kim JE, Chung M-S, Min SC. Cold plasma treatment for the microbiological safety of cabbage, lettuce, and dried figs. Food Microbiology. 2015;**51**:74-80. DOI: 10.1016/j.fm.2015.05.004

[110] Mason TJ, Chemat F, Ashokkumar M. Power ultrasonics for food processing. Power Ultrasonics. Applications of High-Intensity Ultrasound. Cambridge, UK: Woodhead Publishing; 2015:815-843; DOI: 10.1016/b978-1-78242-028-6.00027-2

[111] Cheng X, Zhang M, Xu B, Adhikari B, Sun J. The principles of ultrasound and its application in freezing related processes of food materials: A review. Ultrasonics Sonochemistry. 2015;**27**:576-585. DOI: 10.1016/j.ultsonch.2015.04.015

[112] Guamán-Balcázar MC, Setyaningsih W, Palma M, Barroso CG. Ultrasound-assisted extraction of resveratrol from functional foods: Cookies and jams. Applied Acoustics. 2016;**103**:207-213. DOI: 10.1016/j.apacoust.2015.07.008

[113] Soria AC, Villamiel M. Effect of ultrasound on the technological properties and bioactivity of food: A review. Trends in Food Science & Technology. 2010;**21**(7):323-331. DOI: 10.1016/j.tifs.2010.04.003

[114] FDA. Kinetics of Microbial Inactivation for Alternative Food Processing Technologies—Ultrasound. USA: Food and Drug Administration; 2015

[115] Zinoviadou KG, Galanakis CM, Brnčić M, Grimi N, Boussetta N, Mota MJ, et al. Fruit juice sonication: Implications on food safety and physicochemical and nutritional properties. Food Research International. 2015;**77**:743-752. DOI: 10.1016/j.foodres.2015.05.032

[116] Ozkoc OS, Sumnu G, Sahin S. Recent developments in microwave heating. In: Sun DW, editor. Emerging Technologies for Food Processing. 2nd ed. San Diego: Academic Press; 2014. pp. 361-383. DOI: 10.1016/B978-0-12-411479-1.00020-6

[117] Alarcon-Rojo AD, Janacua H, Rodriguez JC, Paniwnyk L, Mason TJ. Power ultrasound in meat processing. Meat Science. 2015;**107**:86-93. DOI: 10.1016/j.meatsci.2015.04.015

[118] Pingret D, Fabiano-Tixier A-S, Chemat F. Degradation during application of ultrasound in food processing: A review. Food Control. 2013;**31**(2):593-606. DOI: 10.1016/j.foodcont.2012.11.039

[119] Farkas J, Mohácsi-Farkas C. History and future of food irradiation. Trends in Food Science & Technology. 2011;**22**(2–3):121-126. DOI: 10.1016/j.tifs.2010.04.002

[120] Harder MNC, Arthur V, Arthur PB. Irradiation of foods: Processing technology and effects on nutrients: Effect of ionizing radiation on food components. In: Encyclopedia of Food and Health. Oxford: Academic Press; 2016. pp. 476-481

[121] FDA. Irradiation in the production, processing and handling of food. Final rule. Federal register-Food and Drug Administration, HHS. 2012;**73**(164):49593

[122] Diehl JF. Food irradiation—Past, present and future. Radiation Physics and Chemistry. 2002;**63**(3-6):211-215. DOI: 10.1016/s0969-806x(01)00622-3

[123] Morehouse KM, Komolprasert V. Chapter 1—Irradiation of Food and packaging: An overview. Irradiation of Food and Packaging. USA: American Chemical Society; 2004:1-11. DOI: 10.1021/bk-2004-0875.ch001

[124] Kumar S, Saxena S, Verma J, Gautam S. Development of ambient storable meal for calamity victims and other targets employing radiation processing and evaluation of its nutritional, organoleptic, and safety parameters. LWT—Food Science and Technology. 2016;**69**:409-416. DOI: 10.1016/j.lwt.2016.01.059

[125] Marathe SA, Deshpande R, Khamesra A, Ibrahim G, Jamdar SN. Effect of radiation processing on nutritional, functional, sensory and antioxidant properties of red kidney beans. Radiation Physics and Chemistry. 2016;**125**:1-8. DOI: 10.1016/j.radphyschem.2016.03.002

[126] Ravindran R, Jaiswal AK. Wholesomeness and safety aspects of irradiated foods. Food Chemistry. 2019;**285**:363-368. DOI: 10.1016/j.foodchem.2019.02.002

[127] Gautam RK, Nagar V, Shashidhar R. Effect of radiation processing in elimination of *Klebsiella pneumoniae* from food. Radiation Physics and Chemistry. 2015;**115**: 107-111. DOI: 10.1016/j.radphyschem.2015.06.016

[128] Rawson A, Patras A, Tiwari BK, Noci F, Koutchma T, Brunton N. Effect of thermal and non thermal processing technologies on the bioactive content of exotic fruits and their products: Review of recent advances. Food Research International. 2011;**44**(7):1875-1887. DOI: 10.1016/j.foodres.2011.02.053

[129] Caulfeld CD, Cassidy JP, Kelly JP. Effects of gamma irradiation and pasteurization on the nutritive composition of commercially available animal diets. Journal of the American Association for Laboratory Animal Science. 2008;**47**(6):61-66

[130] Cheigh C-I, Park M-H, Chung M-S, Shin J-K, Park Y-S. Comparison of intense pulsed light- and ultraviolet (UVC)-induced cell damage in Listeria monocytogenes and Escherichia coli O157:H7. Food Control. 2012;**25**(2): 654-659. DOI: 10.1016/j.foodcont.2011.11.032

[131] Abida J, Rayees B, Masoodi FA. Pulsed light technology: A novel method for food preservation. International Food Research Journal. 2014;**21**(3): 839-848

[132] Koutchma T, Keener L. Novel food safety technologies emerge in food production. Food Safety Magazine. 2015. Available from: https://www.researchgate.net/publication/272351621_Novel_Food_Safety_Technologies_Emerge_in_Food_Production

[133] Elmnasser N, Guillou S, Leroi F, Orange N, Bakhrouf A, Federighi M. Pulsed-light system as a novel food decontamination technology: A review. Canadian Journal of Microbiology. 2007; **53**(7):813-821. DOI: 10.1139/W07-042

[134] Sizer CE, Balasubramaniam VM. New intervention processes for minimally processed juices. Food Technology. 1999;**53**:64-67

[135] Demirci A, Ngadi MO. Microbial Decontamination in the Food Industry: Novel Methods and Applications. Cambridge, UK: Elsevier, Woodhead Publishing; 2012

[136] Oteiza JM, Giannuzzi L, Zaritzky N. Ultraviolet treatment of orange juice to inactivate E. coli O157: H7 as affected by native microflora. Food and Bioprocess Technology. 2010; **3**(4):603-614. DOI: 10.1007/s11947-009-0194-y

[137] Gabriel AA. Inactivation of Escherichia coli O157: H7 and spoilage yeasts in germicidal UV-C-irradiated and heat-treated clear apple juice. Food Control. 2012;**25**(2):425-432. DOI: 10.1016/j.foodcont.2011.11.011

Section 2

Food Technologies in Food Processing

Chapter 2

Olive Processing: Influence of Some Crucial Phases on the Final Quality of Olive Oil

Lucia Morrone, Annalisa Rotondi, Francesca Rapparini and Gianpaolo Bertazza

Abstract

The extra virgin olive oil (EVOO) chemical and sensory characteristics depend on several factors such as the environment, the genetic matrix, stage of olive ripeness, phytosanitary conditions of olive, time and way of olive storage before transformation, and technological features of olive mill. In this chapter, the time of olive storage and two different types of extraction equipment are taken into account to deep understand their impact on chemical and sensory profile of EVOO. The knowledge of how these factors act will allow to manage the production chain adequately and to act on the various steps in order to improve the quality of EVOO. The sensory modifications of olive oils processed with two different types of extraction system during the storage were also evaluated.

Keywords: extra virgin olive oil, olive storage, oil extraction, Sinolea, decanter, sensory analysis, aroma analysis

1. Introduction

Extra virgin olive oil (EVOO) quality is the result of the interaction of agronomic, pedoclimatic, and technological factors. Among all these factors, the olive fruit characteristics that entering the oil mill are a key factor and probably the most important variable involved in the quality of the final virgin olive oil [1].

The European low stated that EVOO is obtained exclusively through physical procedures as states in European regulation 1513/01 [2], of which the main technological steps are crushing, kneading, and malaxing the oil extraction. Olive oil is one of the few vegetable oils that can be consumed without refining, and so, this makes EVOO comparable to a fruit juice. In fact, EVOO contains phenols responsible for the bitter and pungent taste, and moreover, hydroxytyrosol, a phenolic alcohol, confers health properties as stated by EFSA [3].

In light of this, it is evident that the state of the raw material greatly affects the chemical and sensory characteristics of the EVOO. Moreover, the olive fruit characteristics interact with the technological features of olive mill resulting in different EVOO product's characteristics [4].

Postharvest period of a fruit comprises all the processes that the olive is subjected from harvesting to its industrial transformation. The degree of excellence of virgin olive oil is directly related to the physiological stage of the fruit when processed, and this is the most important factor determining its level of quality [5]. The olive is formed by the epicarp or skin that is composed of 1.5–3.5% of the drupe weight, by the mesocarp or pulp that constitutes between 70 and 80%, by the endocarp or hazel that constitutes between 15 and 25%, and by the almond or seed that has a weight on the total between 2.5 and 4%. The mesocarp is made up mostly of water, oil, and carbohydrates. Triacylglycerols (TGs) are synthesized in plastids and mitochondria of the pulp cell cytoplasm, and then, they merge to produce small oil drops until they reach a diameter of 30 μn. These drops are stabilized by a polysaccharide membrane, unlike what happens in the oil seeds where the oil droplets are incorporated in the oleosomes [6]. The cell wall of the mesocarp cells is rigid and, together with the constituents of the cells, contributes to the firm consistency of the pulp that occurs at the beginning of maturation. During olive ripening, the cell walls become thinner, and the cells are gradually separated due to the solubilization of the pectins and hemicelluloses with consequent softening of the pulp. This phenomenon makes the olive a delicate fruit, whose handling must be done trying to avoid damaging the fruit. The storage of olives in pile, as is often done when there is no synchronization between collection and processing, produces a large heating and crushing of the fruits with a consequent loss of cellular fluids [7]. In these conditions, fruits mechanically damaged are extremely sensitive to fungi infection that leads and accelerates the hydrolytic and oxidative degeneration produced by lipases, lipoxygenases, and liases of both olive and parasitic origin [5]. Fermentative phenomena produce acetic and butyric acids, which cause off flavor in the oil and are responsible for the typical musty smell [7]. Oils produced from these olives have high values of acidity, number of peroxides, and high ultraviolet constants, often above the limits set by regulation 2568/1991 and following amendments that will make them lampante and therefore destined for refining because they are not suitable for human consumption [8]. To prevent this degeneration of the fruits and therefore to avoid obtaining oil with poor chemical and sensorial characteristics, the adoptable strategies are to reduce the storage times by better coordinating the phases of collection and transformation.

Several studies have been conducted to explore the possibility of storing olives in a refrigerated environment. It has been seen that oils obtained from olives stored at 5°C up to 30 days preserved the best characteristics compared to those obtained from olives kept at environment temperature [5, 8]. However, each cultivar can behave differently with respect to both cold storage and storage times [8].

In Emilia Romagna region, one of the northernmost areas in Italy for olive cultivation, the olive harvest phase is well synchronized with the olive mill; however, the olive production in this region is increasing, so a study on the behavior of the storage times of the autochthonous olive cultivars was undertaken. Moreover, a comparison of chemical and sensory characteristics of Nostrana di Brisighella EVOO produced by percolation method, namely Sinolea, and decanter technology was carried out.

2. Impact of the olive storage time on the chemical and sensory characteristics of the oils

Olive oil samples (n = 132) were collected from seven different industrial oil mills located in Emilia Romagna region (Italy). In order to standardize the technological factor, only mills equipped with continuous systems, having hammer crusher, two phase decanters, centrifugation, and filtration phases. Only healthy olive samples without any kind of infection or physical damage were collected.

Results reported in **Table 1** show the analytical determination carried out in accordance with the EU regulation 2568/91 and following amendments. Free acidity showed statistical significant differences only in oils obtained from cultivar mixture, while neither in Nostrana di Brisighella nor in Leccino, a trend was detectable. This indicates the importance of genetic matrix in the deterioration process of oils. The same behavior was shown by peroxide number: only in cultivar Nostrana di Brisighella, a statistical difference was detected. Peroxide number is an indicator of the primary oxidation with a legal limit for EVOO of 20 mEq O_2/kg of oil. K232 and K270 are indexes primarily used to detect frauds, with the legal limits for EVOO of 2.5 for K232 and 0.2 for K270. K232 is also used as an indicator of olive oil primary oxidation, while K270 indicates secondary oxidation in EVOO. Values detected for K232 and K270 were below legal limit but do not discriminate oils according to the time of olive storage.

Phenolic compounds are present in the water dispersion in EVOO. Phenols act as radical scavenging [9], lengthening the EVOO's shelf life. But the long storage times of the olives have led to an impoverishment of the phenolic content of the oils in all samples (**Table 2**). Olive of Nostrana di Brisighella and Leccino stored for 3–6 days showed a decrease in total phenol content and OSI, and a clear reduction trend in both OSI and total phenol content is detectable as the olive storage time proceeds. Olives stored for over 7 days have suffered a drastic breakdown of the phenol content in all oil samples. In particular, the Nostrana di Brisighella oils suffered a phenol loss up to about 76%. This latter cultivar undergoes the phenol degradation in a short time, and probably, its dual purpose attitude makes it delicate. This impoverishment in phenols also affects the stability of the oils. A clear reduction trend was detectable in OSI time in all samples even if in the Nostrana di Brisighella cultivar, the differences were statistically significant. These results agree with studies of Vichi and colleagues [10].

	Time of olive storage	Free acidity[a]	Peroxid number[b]	k232	k270
NdB	<48 h	0.30 ± 0.10	6.47 ± 2.15[a]	1.49 ± 0.58	0.09 ± 0.04
	3–6 days	0.27 ± 0.06	8.03 ± 2.96[b]	1.63 ± 0.7	0.08 ± 0.04
	>7 days	0.28 ± 0.04	9.83 ± 2.35[b]	1.72 ± 0.26	0.08 ± 0.01
	p-Value	0.840	0.046	0.135	0.541
Mix	<48 h	0.33 ± 0.14a	8.62 ± 2.83	1.57 ± 0.38	0.08 ± 0.03
	3–6 days	0.50 ± 0.27b	8.53 ± 2.82	1.56 ± 0.45	0.09 ± 0.03
	>7 days	0.53 ± 0.32b	9.81 ± 3.15	1.67 ± 0.57	0.09 ± 0.03
	p-Value	0.001	0.145	0.152	0.563
Leccino	<48 h	0.33 ± 0.12	7.33 ± 1.97	1.42 ± 0.10	0.06 ± 0.04
	3–6 days	0.37 ± 0.22	7.42 ± 2.96	1.53 ± 0.29	0.08 ± 0.02
	>7 days	0.34 ± 0.1	12.07 ± 1.75	1.79 ± 0.07	0.07 ± 0.01
	p-Value	0.939	0.097	0.223	0.531

The values reported are means ± standard deviation. NdB, Nostrana di Brisighella; Mix, varietal mixture. Different letters in the column indicate significant difference at 5% for each cultivar.
[a]oleic acid in 100 g of oil.
[b]mEq O_2 kg^{-1} g of oil.

Table 1.
Analytical determination of olive oils.

	Time of olive storage	OSI[a]	Total phenols[b]
NdB	<48 h	33.28 ± 9.68[a]	265.31 ± 90.07[a]
	3–6 days	28.06 ± 6.99[a,b]	203.1 ± 123.77[a,b]
	>7 days	15.1 ± 5.37[b]	63.4 ± 23.09[b]
	p-Value	0.020	0.007
Mix	<48 h	23.31 ± 7.93	185.13 ± 79.7
	3–6 days	20.45 ± 7.81	189.3 ± 86.92
	>7 days	19.08 ± 10.07	150.4 ± 68.65
	p-Value	0.092	0.137
Leccino	<48 h	31.21 ± 20.15	251.06 ± 170.76
	3–6 days	19.69 ± 6.69	153.93 ± 78.72
	>7 days	19.43 ± 9.02	108.29 ± 72.41
	p-Value	0.219	0.209

The values reported are means ± standard deviation. NdB, Nostrana di Brisighella; Mix, varietal mixture OSI, Oxidative stability index. Different letters in the same column indicate significant difference at 5% for each cultivar.
[a]*hours.*
[b]*mg of gallic acid kg^{-1} of oil.*

Table 2.
Total phenols content and OSI time detected in olive.

The sensory profile that characterizes an oil is the result of the interaction of numerous substances, both volatile and non-volatile, which stimulate specific receptors allowing us to discriminate the different flavors and smells of olive oil. The oil sensory characteristics are influenced by several factors linked both to the raw material: variety, stage of maturation of the olives, and time and storage conditions and to the extraction technology during which enzymatic reactions take place allowing the formation of aromas [1].

Sensory analysis was performed by the "ASSAM—Marche panel," a fully trained taste panel recognized by the International Olive Oil Council (IOOC) of Madrid, Spain, and by the Ministry for Agriculture, Food, and Forestry Policy.

The sensory profiles of Nostrana di Brisighella olive oil show differences between oils from olive milled within 48 h and after 48 h. In particular, from **Figure 1**, it is possible to see that there is a statistically significant decrease in olive fruity intensity, grass, pungent, and other pleasant notes in oil from olive processed after 48 h. The same trend is detectable in oils from cv. Leccino, of which the radar chart is shown in **Figure 1**. Oils of cv. Leccino milled after the harvest show higher values of all sensory descriptors than oils milled after several days after the harvest. For the cultivar mixtures, influence of the time of storage of the olives was found (**Figure 1**). In fact, differences in olive fruity, grass, bitter, and pungent sensory descriptor were detectable.

However, it is important to underline that in the oil samples with more than 48 h of olive storage time, the percentage of oils with sensory defects was always greater than the oils of the same cultivar with shorter storage time.

With the aim of evaluating the shelf life of the olive oils, the sensory analyses were repeated after 1 year. The EVOO shelf life is a delicate phase since an impoverishment of sensory and chemical characteristics can occur. During the shelf life, oxidation process takes place, and it is characterized by two phases: in the first phase, the oxygen reacts with the unsaturated fatty acids forming hydroperoxides,

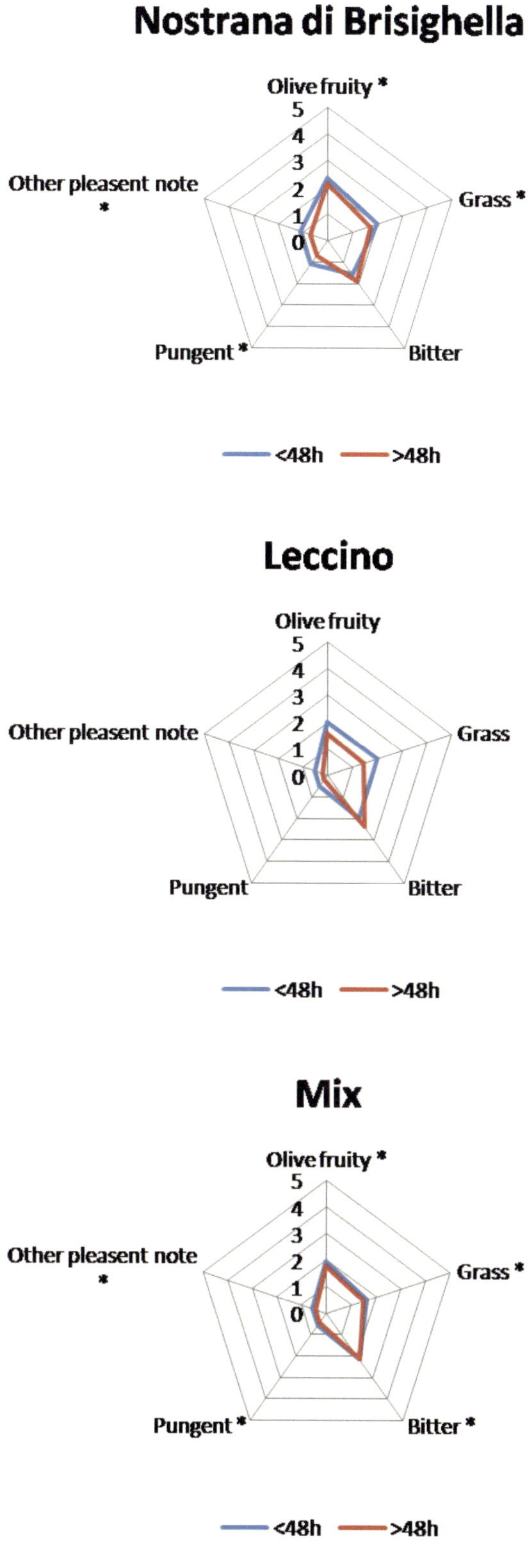

Figure 1.
Sensory profiles of Nostrana di Brisighella, varietal mix, and Leccino processed at different olive storage times (<48 h and >48 h). The asterisks indicate statistical significance at 5% level.

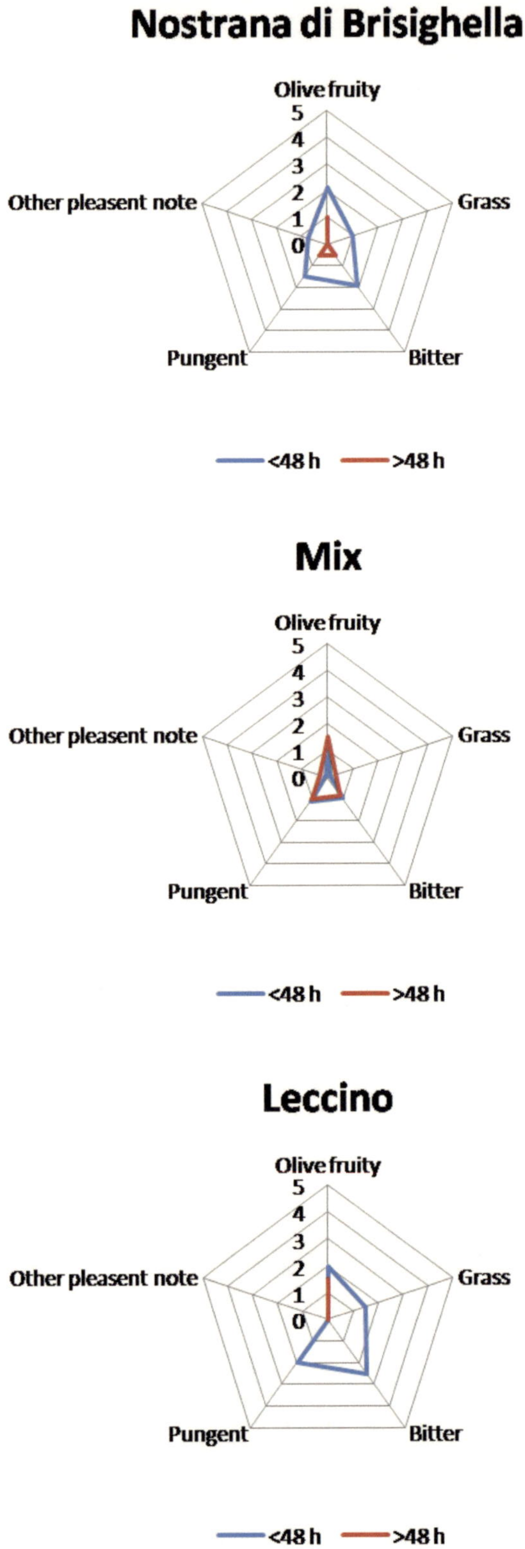

Figure 2.
Sensory profiles of Nostrana di Brisighella, Leccino, and varietal mix after 12 months from oil production (T12).

which, being unstable, fragment itself and give rise to the second oxidation phase that finishes with the formation of ketones and aldehydes. From the radar chart in **Figure 2**, it is possible to see the influence of the olive storage time on the sensory characteristic of oils. In fact, oils produced from olive stored for more than 48 h showed a poor sensory profile compared to olive stored within 48 h. Furthermore, from the comparison of the T12 profiles with those taken just after pressing (**Figure 1**), it is possible to see the greater sensorial degradation of the oils crushed by olives stored for a long time. It is important to underline that oil processed within 48 h maintained their sensory profile over the conservation.

3. Sinolea and decanter: comparison of two extracting methods

The oil extracting method deeply influences the chemical and sensory characteristics of olive oil [4]. In this section, we compare the chemical and sensory characteristics of olive oils obtained using the Sinolea and decanter continuous methods. The Sinolea method exploits the different surface tension of the vegetation water and the oil, and these different physical behaviors allow the olive oil to adhere to a steel plaque, while the other two phases remain behind. It is made up of several metal plates that are dipped into the paste: the oil preferentially wets and sticks to the metal and is removed with scrapers in a continuous process. The decanter centrifugation method exploits centrifugal force allowing the separation of the phases according to their different densities. The study was carried out on the cv. Nostrana di Brisighella. The samples analyzed did not show a significant difference in free acidity and K270, while the peroxide number and K232 revealed differences in the two systems studied (**Table 3**). The peroxide number and K232 give us information about the primary lipid oxidation, so these data suggest a different impact on lipid oxidation of the two extraction methods used. In particular, the Sinolea seems to be more "gentle," and oils extracted using this method were less oxidized.

Tocopherols are lipid soluble vitamins and act as antioxidants by maintaining the cell membrane stability and by preventing the oxidative damage of tissues [11]. Alfa tocopherol has a synergistic effect on ortho-diphenols and contributes significantly to the retardation of peroxide formation [12]. As far as concern the antioxidant substance, the results are presented in **Table 4**. The content of alfa tocopherol was greater in samples extracted with Sinolea than the content of olive oils extracted using decanter. Also, the total phenolic content and the oil stability were greater in oils extracted with Sinolea system. A correlation was found between OSI and phenol content [13], and so, the OSI time is greater in oils extracted using Sinolea than the oils extracted using Decanter.

	Free acidity[1]	Peroxide number[2]	k232	k270
Sinolea	0.28 ± 0.07	5.67 ± 1.36	1.38 ± 0.58	0.08 ± 0.04
Decanter	0.30 ± 0.11	7.16 ± 2.49	1.56 ± 0.59	0.09 ± 0.03
p-Value	0.392	0.027	0.008	0.488

Data are presented as mean ± standard deviation. Different letters in the same column indicate significant difference at 5%.
[1]*g Oleic acid in 100 g oil.*
[2]*Peroxide value, mEq O_2 kg^{-1} of oil.*

Table 3.
Quality indices of virgin olive oil extracted using Sinolea and decanter systems.

	A tocopherol	β + γ tocopherols	OSI	Total phenol
Sinolea	204.97 ± 20.93a	8.5 ± 0.95	34.82 ± 9.23	279.61 ± 66.18
Decanter	185.45 ± 32.23b	8.69 ± 1.47	31.08 ± 10.09	241.37 ± 108.92
p-Value	0.029	0.629	0.203	0.195

Data are presented as mean ± standard deviation. Tocopherol is expressed as mg kg^{-1} of relative standard; OSI is expressed in hours; total phenols are expressed as mg of gallic acid kg^{-1} of oil. Different letters in the same column indicate significant difference at 5%.

Table 4.
Antioxidant fraction of virgin olive oil extracted by two methods: Sinolea and decanter.

The results of sensory analysis of EVOO samples extracted using the Sinolea and decanter systems are shown in **Figure 3**. Oil extracted using Sinolea method presents higher intensities in olive fruity and grass scent than oil extracted using decanter extraction system. These results are in agreement with those of [14] who reported a higher panel score for EVOO extracted using Sinolea than EVOO extracted using decanter.

The sensory analysis was repeated after 12 months in order to verify if the sensory differences detected soon after the EVOO extractions were still present. The result of the sensory analysis carried out after 12 months is shown in **Figure 4**. EVOO extracted using Sinolea had still higher intensities of olive fruity and grass note after 12 months.

It is well known that the production of volatile compounds is a complex process starting when fruit tissues are broken, and enzymes and substrates come into contact [15]. Aside from olive cultivar, geographical origin, fruit ripening degree, and storage conditions, the aroma profile is affected during the fruit processing and oil extraction [16]. We investigated the effect of olive fruit processing and oil extraction using the Sinolea and decanter extraction systems on the volatile content of EVOO of Nostrana di Brisighella cultivar (**Tables 5–7**). The volatile compounds were extracted by dynamic headspace concentration on carbon traps and analyzed by gas chromatography and mass spectrometry. The sampling methodology and the instrument's working parameters for the detection, identification, and quantification of volatiles, were adjusted using the analysis method reported by Rapparini and Rotondi [17] and Vitalini [18]. Briefly, olive oil was extracted with pure He at a rate of 100 ml min^{-1} for 10 min (**Figure 5**).

The headspace volatiles released from 40 ml of oil were collected onto charcoal adsorbent traps (Carbotrap—0.17 g and Carbotrap C—0.034 g; Lara, Rome, Italy). The analytical system consists of a thermal desorber (Chrompack, Middleburg, The Netherlands) connected to a gas chromatograph GC (Hewlett Packard 5890) and a 5970 quadrupole mass spectrometer (MS) as detection system (Hewlett Packard, Palo Alto, CA, USA). All separations are performed on a 60 m × 0.25 mm I.D. capillary column (Hewlett Packard) coated with a 0.25-μm film of polymethylsiloxane. The temperature program was isothermal at 40°C for 7 min and increased to 240°C at 5°C min^{-1}. Identification of the detected compounds is achieved by comparing the retention times, mass spectra of authentic standards (Fluka, Switzerland), and published literature spectra. Quantification of the volatiles was performed when standards were available as previously reported [18]. The individual compound concentrations were calculated by dividing the amount of the volatiles trapped onto the traps by the total sampled air volume and by the total volume of olive oil (ng ml^{-1}).

The combination of dynamic headspace sampling and pre-concentration system with GC-MS analytical technique allowed us to determine in the volatile fraction of

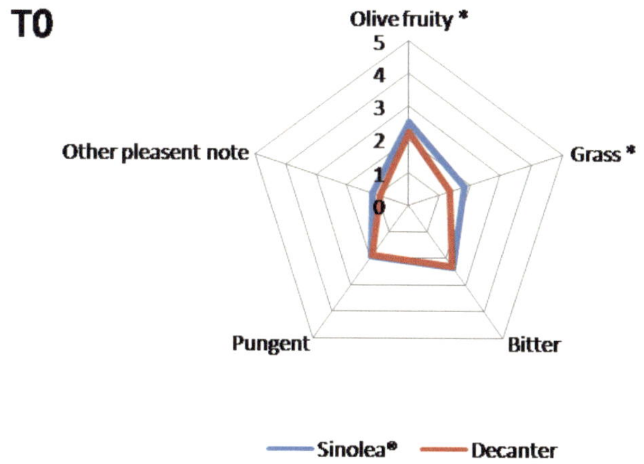

Figure 3.
*Radar chart of sensory intensities indicates from panel test carried out soon after the EVOO production (T0). The asterisk near the sensory attribute indicates a statistical significance difference (Tukey's test; *p < 0.05).*

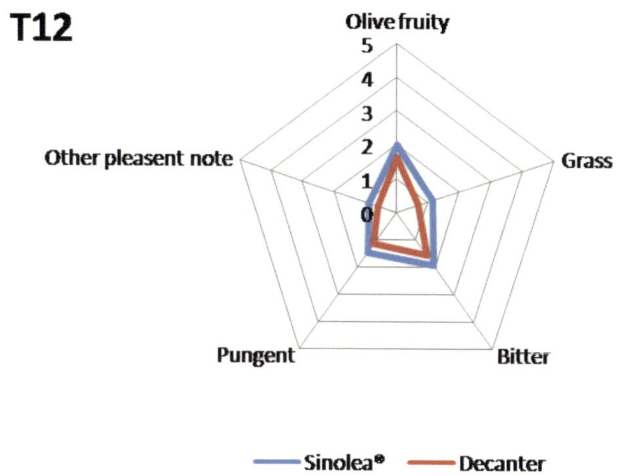

Figure 4.
*Spider chart of sensory intensities indicates from panel test carried out after 12 months of EVOO storage (T12). The asterisk near the sensory attribute indicates a statistical significance difference (Tukey's test; *p < 0.05).*

olive oil samples, a total of 47 compounds (**Table 5**) mainly corresponding to the following chemical classes: alkanes, alcohols, aldehydes, ketones, and esters.

The quantified volatiles are released at a wide range of concentration (from few ng ml^{-1} up to 1911 ng ml^{-1} of oil; **Table 5**). Overall, the total volatile content was higher in olive oil samples of second harvesting campaign (ranging from about 900 to 2500–4000 ng ml^{-1}) than in olive oils obtained during the first campaign (ranging from 200 to 500 ng ml^{-1} of oil), independently of the extraction process employed.

Among the different identified chemical classes, the six-carbon compounds, aldehydes, and alcohols, which have been related to fresh green odor, are especially abundant (**Table 6**). These compounds are produced during the oil extraction by the so-called lipoxygenase (LOX) pathway and activated by the mechanical break of olive fruit [19, 20]. The contribution of the total C6 volatile compounds in the analyzed oils is relevant, representing on average 50% of the total volatiles and reaching a maximum of ca. 72% of the total volatiles in the aroma profile of the oils obtained using the decanter system during the first harvesting campaign (**Table 6**).

Classes	Compounds	
Alkanes	Methyl pentane	X
	Heptane	X
	Octane	X
Alcohols	Ethanol	X
	1-propanol	X
	2-butanol	X
	2-methyl-1-propanol	X
	1-butanol	X
	1-penten-3-ol	X
	3-pentanol	X
	(E)-2-penten-1-ol	X
	(Z)-2-penten-1-ol	X
	3-methyl-1-butanol	X
	2-methyl-1-butanol	X
	(E)-3-hexenol	X
	(Z)-3-hexenol	X
	(E)-2-hexenol	X
	1-hexanol	X
Aldehydes	2-methyl propanal	X
	Butanal	X
	2-butenal	tr
	3-methyl-butanal	X
	2-methyl-butanal	X
	2-methyl-2-butenal	X
	Pentanal	X
	(Z)-2-pentenal	X
	(E)-2-Pentenal	X
	(Z)-3-hexsenal	X
	1-hexanal	X
	(E)-2-hexenal	X
	Benzaldehyde	tr
	Octanal	X
	Nonanal	X
	Ethyl-benzaldehyde	X
	2-nonenal	X
	Decanal	X
	2-Decenal	X
	(E)-2-decenal	X
Ketones	2-Butanone	X
	1-penten-3-one	X

Classes	Compounds	
	2-pentanone	X
	3-pentanone	X
	2-eptanone	tr
	6-methyl-5-epten-2-one	tr
Esters	Ethyl acetate	X
	2-methyl butyl propanoate	X
	Methyl benzoate	X

tr = traces (<0.01 ng/ml of oil).

Table 5.
Volatile compounds detected in the headspace of EVOO of Nostrana di Brisighella cultivar sampled.

Aldehydes are the main fraction of the C6 volatiles, representing about 80–90% of the total C6 compounds from LOX pathway, while C6 alcohols contribute for about 10–13% (**Table 7**). Among the C6 aldehydes, (E)-2-hexenal, which is generally characterized by green, fruity, and almond notes, was the main contributor (72–81%) of the total C6 volatiles (**Table 7**) as usually found for the profile of EVOO [21]. The percentage of the sum of C6 volatiles derived from linolenic acid (LnA) on the total C6 compounds (ca. 83–89%) is in all samples higher than the sum of C6 compounds derived from linoleic acid (LA; 12–17%; **Table 7**), in accordance with previous results on EVOO oils [22]. Other C5 aromatic compounds, mainly ketones (ca. 20–45%) and alcohols (ca. 10%), contribute to the overall aroma profile of Nostrana di Brisighella oils (**Table 5**). As with C6 compounds, LnA-derived C5 volatiles were the major components of the C5 fraction, with 1-penten-3-one and 1-penten-3-ol being the most abundant volatiles among the C5 ketones and C5 alcohol, respectively.

Despite the differences in the absolute concentrations of the volatiles of the EVOO oils obtained during the two different harvesting campaigns, the relative contribution of volatile compounds, which has an high impact on oil sensory quality, is slightly different depending on the oil extraction system.

In particular, when analyzing the volatile composition based on their origin from the LOX pathway, the percentage of the sum of C6 saturated aldehydes and alcohols (i.e., volatiles derived from the LA) results higher in the oils obtained using the Sinolea system (about 15–17% of the total C6 volatiles) than the aroma profile of volatiles from LA of oils extracted using the decanter system (ca. 11–12%; **Table 7**). The aroma profile of the oils obtained using the decanter system is characterized by a higher percentage of the C6 unsaturated volatiles (i.e., volatiles derived from the LnA; 89% of the total C6 volatiles), essentially due to the higher contribution of (E)-2-hexenal, than the relative content of these compounds in the oils from Sinolea (83–85%). Indeed, this volatile was found in greater proportion in the aroma profile of the oil extracted using the decanter system (80% of the total C6 compounds) than in those derived from the Sinolea one (about 72–75%).

Taking into account that alterations of the relative concentrations of volatiles can have a significant impact on the sensorial characteristics of the oil [19, 20], the observed differences, even minor, evidence an impact of the extraction process on the enzymatic production of C5 and C6 volatiles from the LOX pathway, although physicochemical transformations cannot be excluded to be differentially induced by the two employed technological procedure of oil extraction of this cultivar.

Compounds	Decanter		Sinolea	
	1st year	2nd year	1st year	2nd year
2-methyl propanol	3 ± 1	15 ± 2	1 ± 1	10 ± 2
	(1%)	(1%)	(0%)	(1%)
1-penten-3-ol	14 ± 4	110 ± 7	18 ± 1	72 ± 8
	(4%)	(6%)	(6%)	(4%)
3-methyl butanol	3 ± 1	15 ± 1	4 ± 3	8 ± 2
	(1%)	(1%)	(1%)	(0%)
2-methyl butanol	4 ± 1	32 ± 5	2 ± 1	24 ± 9
	(1%)	(1%)	(1%)	(1%)
2-penten-1-ol	2 ± 1	38 ± 4	5 ± 1	29 ± 6
	(1%)	(2%)	(2%)	(2%)
Total C5 alcohols	**25 ± 4**	**210 ± 15**	**31 ± 4**	**143 ± 22**
	(7%)	**(12%)**	**(11%)**	**(8%)**
1-penten-3-one	52 ± 13	629 ± 84	95 ± 9	544 ± 111
	(15%)	(34%)	(33%)	(30%)
2-pentanone	4 ± 2	28 ± 11	2 ± 1	15 ± 3
	(1%)	(1%)	(1%)	(1%)
3-pentanone	12 ± 1	156 ± 14	20 ± 5	103 ± 14
	(4%)	(9%)	(7%)	(6%)
Total C5 ketones	**67 ± 13**	**813 ± 97**	**117 ± 11**	**663 ± 116**
	(20%)	**(45%)**	**(42%)**	**(38%)**
(*Z*)-3-hexenol	17 ± 3	42 ± 10	12 ± 5	78 ± 10
	(5%)	(2%)	(4%)	(4%)
(*E*)-2-hexenol	6 ± 2	24 ± 3	3 ± 1	14 ± 3
	(2%)	(1%)	(1%)	(1%)
1-hexanol	7 ± 2	18 ± 2	3 ± 2	12 ± 2
	(2%)	(1%)	(1%)	(1%)
Total C6 alcohols	**29 ± 6**	**84 ± 14**	**18 ± 5**	**103 ± 14**
	(8%)	**(4%)**	**(6%)**	**(6%)**
Hexanal	24 ± 4	83 ± 18	19 ± 4	133 ± 24
	(7%)	(4%)	(7%)	(8%)
(*E*)-2-hexenal	214 ± 35	712 ± 173	97 ± 9	739 ± 129
	(57%)	(34%)	(34%)	(40%)
Total C6 aldehydes	**238 ± 38**	**796 ± 190**	**117 ± 12**	**872 ± 144**
	(64%)	**(38%)**	**(41%)**	**(48%)**
Total C6 compounds	267 ± 41	880 ± 203	135 ± 16	976 ± 158
	(72%)	(43%)	(47%)	(53%)
Total volatiles	**362 ± 47**	**1911 ± 236**	**283 ± 18**	**1794 ± 270**

Data are expressed as ng ml^{-1} of oil (mean ± standard error). Percentage of the different chemical compound and class relative to the total amount of volatiles is also shown.

Table 6.
Volatile compounds detected in the headspace of the olive oils obtained from Nostrana di Brisighella cultivar and extracted using a decanter or a Sinolea processing system.

	Decanter		Sinolea	
%Compound/sum of C6 compounds	**1st year**	**2nd year**	**1st year**	**2nd year**
(*E*)-2-hexenal	80	81	72	76
3-hexen-1-ol	6	5	9	8
2-hexen-1-ol	2	3	2	1
Hexanal	9	9	15	14
Hexanol	2	2	3	1
C6 aldehydes	89	90	87	89
C6 alcohols	11	10	13	11
Total C6 from LA	12	11	17	15
Total C6 from LnA	89	89	83	85

The percentage of the sum of C6 volatiles derived from linolenic acid (LnA) and from the linoleic acid (LA) on the total C6 compounds is also reported.

Table 7.
Percent distribution of the C6 volatile compounds on the total amount of C6 compounds detected in the headspace of the olive oils obtained from Nostrana di Brisighella cultivar and extracted using a decanter or a Sinolea processing system.

Figure 5.
Dynamic headspace concentration of EVOO aroma compounds.

Although, previous studies on different olive cultivars, including Italian varieties, evidence that aroma profile is strongly genotype-dependent [23], recently, Sánchez-Ortiz and colleagues [15] show a clear influence of the oil extraction process on the formation of several volatiles with a high impact on EVOO's aromatic quality. Volatile compounds could be used as key biochemical markers to improve the oil extraction technology and the related sensory characteristics of the obtained oils.

Therefore, from these data, it is possible to conclude that there are differences in chemical and sensory characteristics in EVOOs extracted using Sinolea and decanter.

4. Conclusion

Chemical and sensory characteristics of EVOO are the result of the interaction of several factors, so in this chapter, we examine the influence of olive storage time. The time between the olive harvest and the transformation has repercussions on the quality analytical indices. These repercussions dependent on olive cultivars: in fact, Nostrana di Brisighella, Leccino, and varietal mixture had different responses in analytical indices. Probably, the difference of the specific cultivar was "silenced" in the mixed variety. Sensory analysis stressed the importance of reduction in the olive storage time before olive transformation. In fact, soon after the oil production, sensory analysis revealed only slight differences in olive oils milled both before and after 48 h. Nevertheless, the sensory analysis repeated after 12 months of oil storage revealed marked differences in the two samples.

In addition, we examine the influence of technological process on the characteristic of EVOO. From the comparison of Nostrana di Brisighella EVOO obtained by Sinolea or decanter equipment, differences in quality index and in tocopherol content were underlined. In particular, EVOO extracted by Sinolea facility had less value of peroxide number and K232 and greater amount of α tocopherol than the EVOOs extracted by decanter. As far as regard, the volatile fraction of EVOO analyzed a total of 47 compounds was found, mainly corresponding to the following chemical classes: alkanes, alcohols, aldehydes, ketones, and esters. Differences in the absolute concentrations of the volatiles of the EVOO oils obtained during the two different crop seasons were observed. The relative contribution of volatile compounds, which has an high impact on oil sensory quality, is slightly different depending on the oil extraction system. In particular, when analyzing the volatile composition based on their origin from the LOX pathway, the percentage of the sum of C6 saturated aldehydes and alcohols (i.e., volatiles derived from the LA) results higher in the oils obtained using the Sinolea system (about 15–17% of the total C6 volatiles) than the aroma profile of volatiles from LA of oils extracted using the decanter system (ca. 11–12%).

Acknowledgements

Research funded by EU Commission, Regulation (EC) No 528/99, in co-operation with the Agriculture Department of Emilia Romagna region, Italy.

Conflict of interest

The authors declare no conflict of interest.

Notes/thanks/other declarations

The authors are grateful to Massimiliano Magli for the statistical analyses and olive growers for furnishing the olive samples.

Author details

Lucia Morrone, Annalisa Rotondi*, Francesca Rapparini and Gianpaolo Bertazza
Institute of Bioeconomy - IBE of the Italian national Research Council,
Bologna, Italy

*Address all correspondence to: annalisa.rotondi@ibe.cnr.it

References

[1] Fregapane G, Salvador MD. Production of superior quality extra virgin olive oil modulating the content and profile of its minor components. Food Research International. 2013;**54**:1907-1914

[2] European Union Commission. Council Regulation(EC) 1513/2001 of 23 July 2001 amending regulation (EC)136/66/EEC and 1638/98 as regards the extension of the period of validity of the aid scheme and the quality strategyfor olive oil. Official Journal of the European Union. 2001;**L201**:4-7

[3] EFSA Panel on Dietetic Products, Nutrition and Allergies (NDA) Scientific opinion on the substantiation of health claims related to polyphenols in olive and protection of LDL particles from oxidative damage (ID 1333, 1638, 1639, 1696, 2865), maintenance of normal blood HDL-cholesterol concentrations (ID 1639), maintenance of normal blood pressure (ID 3781), anti-inflammatory properties (ID 1882), contributes to the upper respiratory tract health (ID 3468), can help to maintain a normal function of gastrointestinal tract (3779), and contributes to body defences against external agents (ID 3467) pursuant to Article 13(1) of Regulation (EC) No 1924/2006. EFSA Journal. 2011;**9**:2033

[4] Morrone L, Pupillo S, Neri L, Bertazza G, Magli M, Rotondi A. Influence of olive ripening degree and crusher typology on chemical and sensory characteristics of Correggiolo virgin olive oil. Journal of the Science of Food and Agriculture. 2017;**97**:1443-1450

[5] García JM, Yousfi K. The postharvest of mill olives. Grasas y Aceites. 2006;**57**:16-24

[6] Murphy DJ. Structure, function and biogenesis of storage lipid bodies and oleosin in plants. Progress in Lipid Research. 1993;**32**:247-280

[7] Olías JM, Garcia JM. Olive. In: Mitra SK, editor. Postharvest Physiological Storage of Tropical and Subtropical Fruits. New York: CAB International; 1997. pp. 229-243

[8] Clodoveo ML, Delcuratolo D, Gomes T, Colelli G. Effect of different temperatures and storage atmospheres on Coratina olive oil quality. Food Chemistry. 2007;**102**:571-576

[9] Visioli F, Galli C. The effect of minor constituents of olive oil on cardiovascular disease: New findings. Nutrition Reviews. 1998;**56**:142-147

[10] Vichi S, Romero A, Gallardo-Chacón J, Tous J, López-Tamames E, Buxaderas S. Volatile phenols in virgin olive oils: Influence of olive variety on their formation during fruits storage. Food Chemistry. 2009;**116**:651-656

[11] Rotondi A, Bertazza G, Magli M. Effect of olive fruits quality on the natural antioxidant compounds in extravirgin olive oil of Emilia-Romagna region. Progress in Nutrition. 2004;**6**:139-145

[12] Blekas G, Boskou D. Antioxidative activity of 3,4-dihydroxyplienylacetic acid and α-tocopherol on the triglyceride matrix of olive oil. Effect of acidity. Grasas y Aceites. 1998;**49**:34-37

[13] Rotondi A, Bendini A, Cerretani L, Mari M, Lercker G, Gallina Toschi T. Effect of olive ripening degree on the oxidative stability and organoleptic properties of cv. Nostrana di Brisighella extra virgin olive oil. Journal of Agricultural and Food Chemistry. 2004;**52**:3649-3654

[14] Ranalli A, Ferrante ML, De Mattia G, Costantini N. Analytical

evaluation of virgin olive oil of first and second extraction. Journal of Agricultural and Food Chemistry. 1999;**47**:417-424

[15] Sánchez-Ortiz A, Aymen Bejaoui M, Quintero-Flores A, Jiménez A, Beltrán G. Biosynthesis of volatile compounds by hydroperoxide lyase enzymatic activity during virgin olive oil extraction process. Food Research International. 2018;**111**:220-228. DOI: 10.1016/j.foodres.2018.05.024

[16] Angerosa F, Mostallino R, Basti C, Vito R, Serraiocco A. Virgin olive oil differentiation in relation to extraction methodologies. Journal of the Science of Food and Agriculture. 2000;**80**:2190-2195

[17] Rapparini F, Rotondi A. Volatile compounds analysis in virgin oil by dynamic headspace method, preliminary results. Acta Horticulturae. 2002;**586**:695-697

[18] Vitalini S, Ruggiero A, Rapparini F, Neri L, Tonni M, Iriti M. The application of chitosan and benzothiadiazole in vineyard (*Vitis vinifera* L. cv. *Groppello gentile*) changes the aromatic profile and sensory attributes of wine. Food Chemistry. 2014;**162**:192-205. DOI: 10.1016/j.foodchem.2014.04.040

[19] Morales BMT, Angerosa F, Aparicio R, De I, Avda G, García P. Effect of the extraction conditions of virgin olive oil on the lipoxygenase cascade: Chemical and sensory implications. Grasas y Aceites. 1999;**50**:114-121

[20] Sánchez-Ortiz A, Pérez AG, Sanz C. Synthesis of aroma compounds of virgin olive oil: Significance of the cleavage of polyunsaturated fatty acid hydroperoxides during the oil extraction process. Food Research International. 2013;**54**:1972-1978. DOI: 10.1016/j.foodres.2013.03.045

[21] Angerosa F. Influence of volatile compounds on virgin olive oil quality evaluated by analytical approaches and sensor panels. European Journal of Lipid Science and Technology. 2002;**104**:639-660

[22] Vichi S, Castellote AI, Pizzale L, Conte LS, Buxaderas S, Lopez-Tamames E. Analysis of virgin olive oil volatile compounds by headspace solid-phase microextraction coupled to gas chromatography with mass spectrometric and flame ionization detection. Journal of Chromatography A. 2003;**983**:19-33

[23] Angerosa F, Basti C. The volatile composition of samples from the blend of monovarietal olive oils and from the processing of mixtures of olive fruits. European Journal of Lipid Science and Technology. 2003;**105**:327-332

Chapter 3

Powder Technology

Suzana Caetano da Silva Lannes and Maria Elena Del Dolores Bernal Gómez

Abstract

Combining two or more granular or powder ingredients requires a suitable mixing process, which can be either free or random flow with no attraction forces between the particles or interactive or orderly with the presence of large active particles that attract others forming stable clumps. Food systems have very complex properties that make it difficult to standardize the mixing process. In order to achieve an efficient mixture, diffusive and convective mechanisms must be combined, and its success is achieved with a predominance of homogenization over segregation. Powder products are typically used in industry as dispersion in a liquid and should have some properties such as good wettability, water incorporation, flowability, and instantization. To work with powder products, it is necessary to make determinations such as density, particle size, texture, and compaction force, among others. All these physical properties affect and determine the behavior of powdered products during storage, handling, and processing.

Keywords: physical properties, powder products, solid particles, mix process

1. Introduction

Power mixing involves several steps. The first one could be mentioned as a classification of the powder particles. Flowability can be the result of a good classification step. Consistent feed from bulk storage containers into feed mechanisms of subsequent processing operations is necessary. Interparticle forces, including van der Waals forces, capillary, and electrostatic forces influence the behavior of powder flow systems, as well as a small amount of moisture. Flow properties as angle of internal friction related to cohesion force of solids are also determined.

Powder mixing requires a statistical methodology before choosing the right equipment. Only in this way is a satisfactory result obtained, with the distribution of the mixture components as close as possible to ideality.

Segregation tends to occur whenever bulk material moves, and it occurs where differential forces act on different fractions of the mass of bulk material, and when differences in particulate properties cause a preferential movement of particles.

The quality of mix and characterizing it requires taking several samples and analyzing them, as a random way. Measurement of the mixing profile in real time with near infrared (NIR) provides the opportunity to study the dynamics of powder mixing and enabling a more comprehensive statistical analysis [1].

This chapter aims to present some aspects of this powder technology.

2. Sieving

Before starting a powder mixing process, a classification and maybe separation of the particles is necessary.

Sieving process is the separation of a mixture of grains of different sizes in two or more plots, through a sieving surface, which acts as a gauge that allows and does not let the grains pass. The final plots consist of more uniformly sized grain than the original blend. Mesh is the number of apertures of a screen of the same dimension in each linear inch, counted from the center of any wire to a point exactly 1 in (25.4 mm), or by a specified aperture in inches or millimeters, which is understood to be the free opening or space between the wires. Example: A granular material (−10 + 100) means that everything passes through a 10-mesh sieve (particles smaller than 1.68 mm) and nothing passes through a 100 mesh (particles larger than 0.149 mm). Screen opening is the minimum clear space between the edges of the openings on the sieving surface, given in inches or mm.

Particle size distribution is the relative percentage by weight of the grains that constitute the different size fractions present in the sample. It is one of the most important factors in evaluating the screening operation and is best determined by a full-size analysis using test sieves.

3. Powder mix

Mixture can be defined as the result of combining two or more ingredients. It can be granular or powdery. For such granular or powder mixtures to be formed a suitable mixing process is required. According to Pernenkil and Cooney [2], powder mixing is a crucial unit operation in the food industry.

Mixing is considered as a critical factor, especially in case of strong drugs and low dose drugs where high amounts of adjuvants are added.

There are two types of mixtures, non-interactive or random, and interactive or ordered. The first are those of free flow, being mixtures of uniform particle size powders or grains, without intraparticle forces of attraction, thus flowing with little interruption. Consequently, each different particle will have the same probability of being found in any portion of the mixture. Interactive mixing is formed when large active surface particles exist where other particles are attracted. They form stable clusters and the force between the particles belongs to different chemical classes [3].

According to Fellows [4], it is not possible to obtain a completely uniform mixture of powder products or particulate solids, but according to Singh and Heldman [5], the most important fact in a mixture is the reduction of random mix variation.

There are basically three mechanisms in mixing solids: diffusion, convection and shear. Shear can be considered as convective, and efficient mixing must be combined by diffusive and convective mechanisms. A purely diffusion process generates high efficiency in the mixing of individual particles, however, occurs at a low rate. The basically convective process is fast but less effective, exhibiting an ineffective final blend. For solids the diffusive mixture will only occur by mechanical agitation. The particles will change their collective or individual relative positions, and segregation of the particles may also occur, occurring when particles of different sizes, shapes or densities are mixed. A good mix occurs when there is homogeneity of the particles.

It is difficult to define and evaluate the powder mix; but certain quantitative measurements in solids can help estimate mixer performance. The proof of the mixer in practice comes from the properties it provides to the final blend produced by it (**Figure 1**).

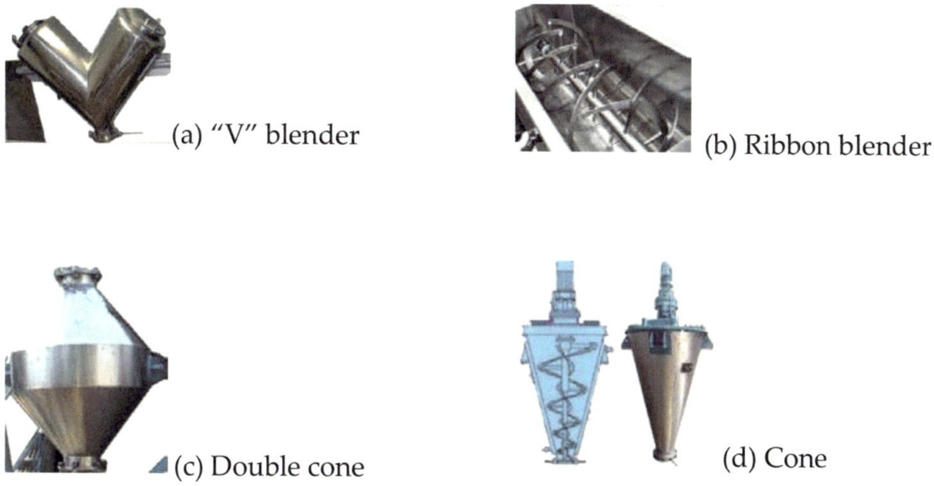

Figure 1.
Mixers.

The design of the mixer and its operation must be carefully chosen to achieve the desired results [4], as this influences the final product quality [6].

Mixing index involves the comparison of standard deviation of sample of a mixture under study with the estimated standard deviation of a completely random mixture (Eq. 1).

$$M_1 = \frac{\sigma_m - \sigma_\infty}{\sigma_0 - \sigma_\infty} \; \sigma_0 - \sqrt{[V_1(1 - V_1)]} \tag{1}$$

where σ_∞ = the standard deviation of a 'perfectly mixed' sample, σ_0 = the standard deviation of a sample at the start of mixing and σ_m = the standard deviation of a sample.

V = the average fractional volume or mass of a component in the mixture [4].

Due to the complexity of the properties of food systems, which may vary during the mixing process, it is extremely difficult to generalize or standardize the mixing operation for various new or traditional applications. The development of mathematical modeling for the food mixing process is also scarce, and it is necessary to consult established procedures for equipment design or scaling up [3].

Near-Infrared (NIR) spectroscopy can be used of in-situ as the basis for an in-line control system to optimize mixing time of food powder blends [7].

Some recommendations before starting the process: Determine which particle properties are required to solve your problem; For the form it is necessary to use an image analyzer. If not, assume that the measured size is an equivalent spherical diameter; for many non-spherical particles do not try different techniques for size checking. Many particles are globular enough to be considered spherical in order to do a job.

4. Physical properties

Powder products have different physical properties that must be measured and studied to obtain a product with the desired characteristics.

Detailed information on the physical properties of powder products is required, especially as they are complex products [8].

Some forces acting on the particles, as Van der Waals, electrostatic, and surfaces forces. Cohesive forces and frictional forces result in surface-surface interactions which resist the movement of particles, and they should be minimal. During mixing, the particles develop surface charge, which produces repulsions between particles, occurring random mixing, depending on surface properties, polarity, charge, and moisture.

Normally, powder products are used in industry as dispersion in a liquid. The wettability test is a simple test used by industry that provides the time parameter required for the powder to be absorbed by a liquid. Although maximum product wetting time is an arbitrary choice, powders in which 90% of the sample has already been dipped within 5 minutes are good wetting [9].

Powder flowability is defined as the ease with which a powder will flow under a specified set of conditions. Some of these conditions include the pressure on the powder, the humidity of the air around the powder and the equipment the powder is flowing through or from. Quantify powder flow characteristics are Compaction, Cohesion, Compressibility and Bulk Density. Flowability cannot showed as a single value or index, due to the combinations of physical properties of materials, the used equipment and processing.

Some physical properties of the powders such as angle of repose or rest angle are of importance for information on product flowability. During powder reconstitution, surface moisturizing water molecules tend to reduce inter-particle cohesiveness, thus allowing faster water penetration, so powders with high angle of repose have greater difficulty in incorporating water [10]. Powders with an angle of repose of up to 40° usually flow easily, if the angle exceeds 50° the flowability may be impaired indicating lower flowability. Particulate solids with up to 35° angle of repose have good flowability, those of 35 - 45° have poor cohesiveness, those of 45 - 55° have good cohesiveness and those above 55° are very cohesive, with low cohesive cocoa powder (45°) and cohesive (52°) cupuassu powder, for example [11]. Some powders show changes in fluidity with storage time [12].

The settled density of powders can be easily determined with a graduated cylinder (20 g sample) with some stirring to constant volume [12]. For example, cupuassu powder has 0.53 g/mL and cocoa powder 0.51 g/mL [11]. Shittu and Lawal [10] analyzed commercial chocolate and found values ranging from 0.49 to 0.81 g/cm^3. Eduardo [13] found values ranging from 0.28 to 0.94 g/cm^3 for chocolate drink powder from market.

Some powder properties:

1. Wettability

Time required for a specific amount of dust to be completely wetted when it is placed in water at a specific temperature. It is mainly related to particle size and shape, temperature and liquid type. Particle surface characteristics and fat content and characteristics if present and the correlation between wettability and fat content are inverse. Important analysis for powder products that will undergo the reconstitution process, as from the wettability analysis can obtain information about the product, such as its dispersibility and tendency to agglomerate formation. The wettability test is used by industry and it is the most important step in the process of reconstituting powder products.

Within this physical property of powders, there are some forms of measurement such as immersion, capillary rise, condensation and spreading. Immersion is the traditional method, which is used in powders that wet reasonably well [14].

2. Solubility index

Determines the ability of the powder to dissolve in water. It is defined as the volume of sediment in mL after centrifugation. The powder is dissolved in water at a certain temperature and centrifuged. The supernatant is removed and replaced with water and centrifuged before reading the volume of insoluble residue.

3. Bulk/tapped density

It is the weight of the powder divided by the volume occupied, usually expressed in g/mL. The sample is placed in an aluminum cylinder, heavy and beaten (100 or 1250 times).

4. Particle size distribution

Particle size is a determining parameter in the effectiveness of homogeneity in a powder mix when these are mixtures of two or more components of different particle sizes [15].

Sieves can be used. The dust sample is divided into fractions with different particle sizes by sieving, or by Laiser (Mastersizer Malvern Equipment).

Particle size distribution and particle size are of utmost importance when studying powder products. The particle size distribution can be represented graphically by the accumulated relative frequency (usually given as a percentage) or by size frequency histograms at certain intervals. It should be considered in the analysis that more than 20% of the material cannot be retained in the first sieve or bottom, and more than 30% of the material cannot be retained in any intermediate sieve [16].

The physical properties of powdered products affect their behavior during storage, handling and processing. Therefore, the determination of such parameters is of great importance for industries that use powders as raw material or even as final product. This is the case in the building materials, ceramics, pharmaceuticals and food industries, among others [17].

Properties of the ingredients of a match that affect the mixing of solids:

1. Particle size distribution—reports the material fractions at different size ranges.

2. Bulk density—weight per unit volume of solid particles. It is not a constant. It can be diminished by aeration and increased by vibration or mechanical compaction.

3. Particle shape—ovoids, blocks, spheres, flakes, chips, rods, filaments, crystals, irregular shapes.

4. Surface characteristics - surface area and the tendency to retain electrical charge.

5. Flow characteristics—rest angle and flowability. They are measurable characteristics determined in standardized assays. A higher rest angle indicates lower flowability. An object resting on an inclined plane begins to slide when the inclination angle is increased sufficiently to overlap the frictional force between the object and the plane. In general, if the angle exceeds 50° the powder will not flow satisfactorily. With about 25° will flow easily (**Figure 2**).

6. Reliability—is the tendency of the material to break during the handling operation. One should also consider the abrasion between the ingredients.

7. Agglomeration state—refers to the independent existence of particles or their adherence to each other, forming aggregates. The type and amount of energy employed during mixing and the friability of the agglomerates will influence aggregate breakdown and particle dispersion.

8. Moisture or liquid content in the solid—often a small amount of liquid is added to the solid to reduce dust or satisfy a special need.

9. Viscosity and surface tension—at the operating temperature of any added liquid.

10. Thermal Limitations of Ingredients—Any effect caused by temperature change must be observed.

Segregation mechanisms can occur with poor flow properties, particle size difference, difference in mobilities and in particle density and shape, transporting methods, dusting stage. Can be summarized as:

1. Fine particle percolation. If a particle mass is disturbed such that individual particles move, a rearrangement of the particle packing occurs;

2. Increase of coarse particles in vibration.

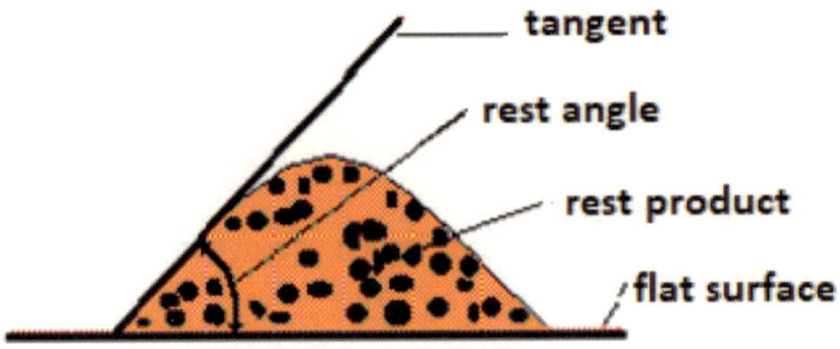

Figure 2.
Rest angle determination.

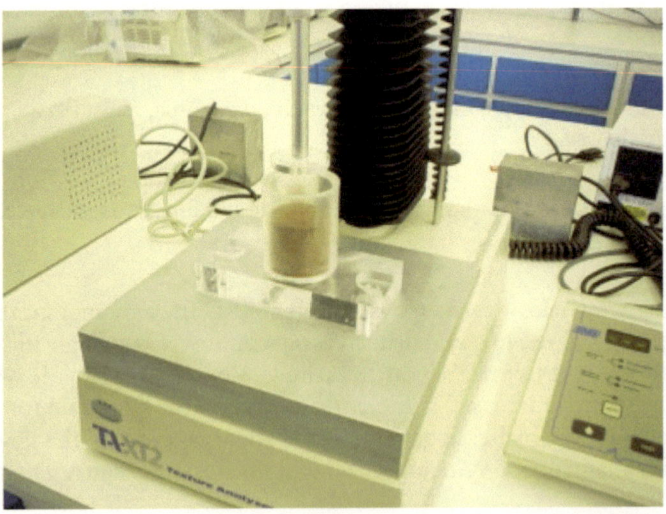

Figure 3.
Texturometer, back extrusion *probe and cylindrical cup with sample.*

4.1 Compression

Compaction can be understood as the compression of a two-phase system, solid and gas (dust and air), under the action of a force, which results in a reduction in the volume of the product. Compaction determination is useful for flow evaluation, friction tendency and dust agglomeration. In industry, the compaction process is used when forming powders, such as tablets. Under a compressive force, the particles rearrange (increasing the density of the dust), deform, and fragment [11, 13, 18].

However, these transformations continue to happen even when compression is not desired. Cartwright [19] associated the dispersibility of powders with their texture. He stated that very fine particles should be avoided when a good instantaneous powder is desired.

Eduardo and Lannes [16] developed a methodology for determining the compaction strength of powders using the TA-XT2 texturometer and the *back extrusion* probe. Medeiros [9] complemented with the compression distance test, aiming to determine the maximum volume reduction occupied by the sample, but that would not exceed the 20,000 g force. The compression strength test is performed at the distances determined in the first test (**Figure 3**).

The relationship between the compaction force and the compaction capacity of the sample is inversely proportional; hence, a sample is most compactable if its compaction force is lower. Based on these data, the most compactable sample was cocoa, as it presented the lowest compaction force, and the least compactable was cupuassu powder, there was no significant difference between them [16].

Eduardo and Lannes [13, 16] determined the compaction force of commercial chocolate, the results obtained ranged from 532 to 16,399 g. From these results the chocolate products were classified as very compact, with force below 2000 g and little compact, with force above 2000 g. These results also depend on the intrinsic characteristics of the particles, such as shape, size and homogeneity.

In granular materials (such as powders) pressure can cause permanent volume change. The removal of air between particles, causing a change in dust volume, can be caused throughout the storage period, transportation or even processing if some type of vibration is involved. Powder products contain in their formulation a great diversity of ingredients with distinct particle characteristics, and the reduction in volume is due to the accommodation of smaller particles between the space left by the larger ones (particle percolation).

5. Instantization

The instantization property identifies foods that are easy to solubilize in cold water, obtained in the drying process using dispersing substances, or through the action of agglomeration [9, 20].

Several physical and chemical methods have been employed to improve the instant properties of powdered foods, as is the case of adding cereal alcohol with its subsequent evaporation under controlled time, temperature and relative humidity, as showed by Barros [21].

One of the methods used to achieve instantization of powdered products is the spray-drying procedure that atomizes a solution by hot air [9, 22].

Spray-drying technology is widely used in various industrial segments including pharmaceutical and food. Although it is a technology that requires large investments in facilities and operation, there are many reasons why it is widely used. These advantages include consistent quality particle production, continuous use, the applicability of the technique to thermosensitive and heat resistant materials, the ability to process various types of raw materials, and the flexibility to define a project based in the formulation. To make use of these advantages, there are several aspects that must be considered. These include the evaluation of the formulation and process parameters, the specific type of particle to be produced and the properties of the material used [9, 23].

Mist drying is the transformation of low or high viscosity liquids, even those that are almost pasty, into dry and pulverized product in a single operation. The liquid or paste is atomized using a centrifugal or high-pressure system where the atomized droplets immediately meet a hot air flow. The rapid evaporation allows keeping the temperature of the product low. Heat and mass transfer are accomplished by direct contact between the hot gas and the dispersed droplets. Fine particles are separated from gas in external cyclones or collecting sleeves. When only the coarse fraction of the finished product is desired, the fines can be recovered in washers; washer liquid is concentrated and returned to the dryer [24–26].

The main use of spray dryers is the drying of solutions and aqueous suspensions. They are also used in combined drying and heat treatment operations. Feed is usually a liquid solution, suspension or paste that can be sprayed [9, 27]. The product to be dried goes through nozzles of varying sizes, influencing the particle size obtained, the liquid part is transformed into an atomized spray. The dust is carried in an airstream that carries it in contact with the spray.

Improvement of the physical and chemical characteristics of the materials used in this technique generally involves the comparison of process parameters such as heating, air volume, atomizer nozzle type, flow rate of the material to be dried or atomization system, drying air temperature. Formulation parameters are evaluated together with process parameters. It is important to check, for example, when the temperature is raised, if there is no extensive protein denaturation, loss of flavor, as well as impairment of solubility, stability and compaction [23, 28].

Heating and mass transfer during drying occur with air and vapor films around the droplets. This vapor shield keeps the particle at saturation temperature. As the particle does not become dry, evaporation continues, and the temperature of the solids does not approach the temperature of the drying outlet. Because of this, sensitive products can be dried at relatively high temperatures.

The shape of most atomized particles is spherical, which ensures fluid-like flow. This helps in the handling and filling process, for example, as well as in reducing costs. The particles still have homogeneity in the composition and the particle size distribution is very close, minimizing the obtaining of very fine particles, which is very important for the obtained product.

Factors such as humidity and water activity are of great importance in the study of the obtained product. Process definition and suitability of equipment operating parameters are particular to each desired finished product, depending on the characteristics it is intended to provide [29].

As the spray-dryer technique is widely used in the industry, the study of its potentiality and suitability to obtain powder products is a way to study the drying process for this and other products, as well as explore the equipment and its resources, obtaining a differentiated product. Instantization and improving product wettability are very important factors in obtaining a powder product, where the drying technique becomes a means to obtain these characteristics. Straatsma et al. [30] studied the solubility index of spray-dryer instantized materials, and this index is of primary importance for instantized powders. The thermal load of food products during drying is an important factor in the final quality of the powder, since heat exposure can lead to the formation of insoluble materials which are undesirable especially for instant powders. Spray-dryer equipment can be seen in **Figure 4**.

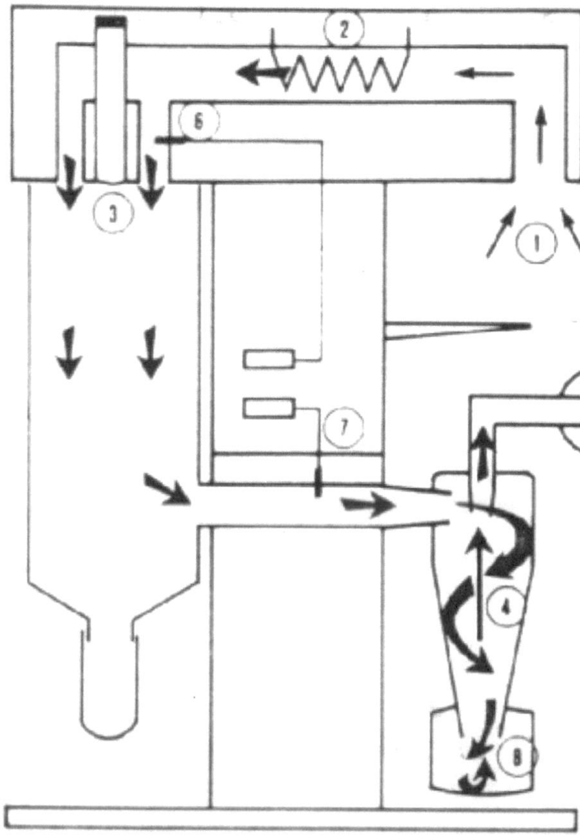

Figure 4.
Scheme of spray-dryer and drying air flow [31] 1. Air inlet, 2. Heating, 3. Entering the drying chamber, 4. Cyclone, 5. Vacuum Cleaner, 6. Control of inlet air temperature, 7. Control of outlet air temperature and 8. Receiving vessel of final product.

Optimal selection of inlet and outlet temperature differences is one of the most important aspects of spray-dryer. The outlet temperature cannot be chosen as desired as it results from the combination of inlet temperature - vacuum adjustment and product feed pump performance.

Product feeding and the introduction of the drying air in this type of dryer are performed at the top of the chamber in co-current flow system. Drying takes place while hot air and the product in the form of small droplets travel through the drying chamber to its conical base. Moist air and dry product then follow to the cyclone, where they are separated, and moist air is removed, and the dry powder product is collected at the base of the cyclone.

Atomizing a powder mixture involves a combination of ingredients, improving wettability in water or another liquid, evening out powder particles as well as improving their flowability and dispersibility. Its high cost must be offset by these factors [32].

Dispersibility is the ability of the powder concentrate to suspend in water to form finely divided particles that will remain in suspension for a reasonable period. It is described as a carrier surface feature and dispersing agents are added by overlapping the forces of attraction between the particles [33]. The dispersion of solids is affected by the texture of the powder, and to be instantaneous the powder must be optimal in size and very fine particles should be avoided [19]. Proper formulation requires a balance between aggregate size and interactions between different chemical additives, as well as adjustment of grinding process conditions.

The degree of atomization influences the drying rate, as well as the residence time of the particles influences the drying size. All atomization techniques can provide good control over the average particle size, but there are differences in their distribution [27].

The concentration of the input product in the atomizer influences the particle size, higher concentration of the solution, the more porous the particles obtained. The lowest concentration provides the smallest and finest particles. Higher flow of atomized product leads to smaller particles in the final product [19]. The adjustment of the process parameters, formulation, atomized product concentration, temperatures, spray speed, should aim at higher yield.

The quality of powdered foods is based on the properties variety that depend on specific applications. In general, final moisture content, solubility, rheological properties of the powder and density are of prime importance. Currently the main challenges in powder production are product development and process cost reduction. As a result, the production capacity is maximized, process conditions are directed to minimal product losses, reduced energy consumption, online quality control [29, 33].

Spray drying is nowadays a technology widely used in the food industry. The purpose is to protect thermosensitive active substances. Many researches are being developed using the microencapsulation method. Thus, to protect oils from lipid oxidation [34–36], incorporating functional ingredients such as vitamins [37, 38], additives and their storage protection [38], antioxidant protection [39].

6. Moisture sorption isotherms

Knowledge of sorption isotherms of powders blend is important for generating data for storage procedures such as shelf life prediction as well as drying processes when this is used in the process. Sorption isotherm can be defined as the graphical representation of the relationship between different humidity and water activity (parameter that describes the degree of binding of water to food particles) at constant temperature [40, 41].

In food, the microscopic structure is of primary importance in all aspects of its functionality. The microscopic organization of both water and other components determines the outcome of macroscopic observations made using different techniques [42–44].

Water activity of a product is defined as the ratio of water vapor pressure to pure water vapor pressure at the same temperature, and the availability of water-based criteria that can provide indicators of stability include water content, solute concentration and osmotic pressure.

7. Final considerations

A perfect mixture of two or more types of solid particles is one in which a sample contains the same proportion of components as any part of the mixture. Mixing of powders is a process that involves a comprehension of the physical elements of the mixture, equipment design, and appropriate sampling technique to ensure mix quality.

DOI: http://dx.doi.org/10.5772/intechopen.90715

Author details

Suzana Caetano da Silva Lannes* and Maria Elena Del Dolores Bernal Gómez
Department of Biochemical-Pharmaceutical Technology, Pharmaceutical Sciences School, University of São Paulo - USP, São Paulo, SP, Brazil

*Address all correspondence to: scslan@usp.br

References

[1] Zanzoni A, Montecchi-Palazzi L, Quondam MX. Mint: A molecular interaction database. FEBS Letters. 2002;**513**:135-140. DOI: 10.1016/s0014-5793(01)03293-8

[2] Pernenkil L, Cooney CL. A review on the continuous blending of powders. Chemical Engineering Science. 2006;**61**(2):720-742

[3] Onwulata C. Encapsulated and Powdered Foods. 1st ed. Boca Raton: CRC Press; 2005. p. 528

[4] Fellows PJ. Tecnologia do processamento de alimentos: Princípios e prática. 2.ed. Artmed: Porto Alegre; 2006. p. 602

[5] Singh RP, Heldman DR. Introduction to Food Engineering. 4th ed. Amsterdam: Academic Press; 2009. p. 841. (Food science and technology international series)

[6] Laurent BF, Bridgwater J. Influence of agitator design on powder flow. Chemical Engineering Science. 2002;**57**(18):3781-3793

[7] Barone A, Glassey J, Montague G. Towards online near-infrared spectroscopy to optimize food product mixing. Journal of Food Engineering. 2019;**263**:227-236

[8] Dhanalakshmi K, Ghosal S, Bhattacharya S. Agglomeration of food powder and applications. Critical Reviews in Food Science and Nutrition. 2011;**51**(5):432-441

[9] Lannes SCS, Medeiros ML. Processamento de achocolatado de cupuaçu por spray-dryer. Revista Brasileira de Ciências Farmacêuticas. 2003;**39**(1):115-123

[10] Shittu TA, Lawal MO. Factors affecting instant properties of powdered cocoa beverages. Food Chemistry. 2007;**100**(1):91-98

[11] Medeiros ML. Estudo e aplicação de substitutos de cacau [thesis]. São Paulo: Pharmaceutical Sciences School-University of Sao Paulo; 2006

[12] Teunou E, Fitzpatrick JJ. Effect of relative humidity and temperature on food powder flowability. Journal of Food Engineering. 1999;**42**(2):109-116

[13] Eduardo MF. Avaliação reológica e físico-química de achocolatados e bebidas achocolatadas [thesis]. São Paulo: Pharmaceutical Sciences School-University of Sao Paulo; 2005

[14] Ji J, Fitzpatrick J, Cronin K, Crean A, Miao S. Assessment of measurement characteristics for rehydration of milk protein based powders. Food Hydrocolloids. 2016;**54**:151-161

[15] Shenoy P, Viauc M, Tammel K, Innings F, Fitzpatrick J, Ahrné L. Effect of powder densities, particle size and shape on mixture quality of binary food powder mixtures. Powder Technology. 2015;**272**:165-172

[16] Eduardo MF, Lannes SCS. Use of texture analysis to determine compaction force of powders. Journal of Food Engineering. 2007;**80**:568-572. DOI: 10.1016/j.jfoodeng.2006.06.011

[17] Fitzpatrick JJ, Iqbal T, Delaney C, Twomey T, Keogh MK. Effect of powder properties and storage conditions on the flowability of milk powders with different fat contents. Journal of Food Engineering. 2004;**64**(4):435-444

[18] N'dri-Stempfer B, Oulahna D, Eterradossi O, Dodds JA. The incidence of pressure on the colour of compacted powders. Powder Technology. 2002;**128**(2-3):320-325

[19] Cartwright B. Spraying provides instant solutions. Food Flavourings, ingredients, Packaging & Processing. 1981;**3**(8):29-33

[20] Vissotto FZ, Montenegro FM, Santos JM, Olieira SJR. Avaliação da influência dos processos de lecitinação e de aglomeração nas propriedades físicas de achocolatado em pó. Ciência e Tecnologia de Alimentos. 2006;**26**(3):666-671

[21] Barros DJM. Desenho e avaliação de formulações de achocolatados obtidos por processos convencional e instantâneo [thesis]. São Paulo: Pharmaceutical Sciences School-University of Sao Paulo; 2016

[22] Turchiuli C, Gianfrancesco A, Palzer S, Dumoulin E. Evolution of particle properties during spray drying in relation with stickiness and agglomeration control. Powder Technology. 2011;**208**(2):433-440

[23] Wendel S, Celik M. Uma visão geral sobre o uso da tenologia de spray-drying. Pharmaceutical Technology. 1998;**2**(2):129-134

[24] Masters KS. Spray Drying in Practice. SprayDryConsult International ApS: Charlottenlund; 2002

[25] Sormoli ME, Langrish TAG. Moisture sorption isotherms and net isosteric heat of sorption for spray-dried pure orange juice powder. LWT-Food Science and Technology. 2015;**62**:875-882

[26] Van't Land CM. Industrial Drying Equipment. Selection and Application. New York: Marcel Dekker; 1991. p. 362

[27] Perry RH, Chilton CH. Manual de Engenharia Química. 5th ed. Rio de Janeiro: Guanabara Dois; 1980

[28] Rhodes MJ. Principles of Powder Technology. New York: Wiley; 1990. p. 439

[29] Straatsma J, Van Houwelingen G, Steenbergen AE, De Jong P. Spray drying of food products: 1. Simulation model. Journal of Food Engineering. 1999a;**42**:67-72

[30] Straatsma J, Van Houwelingen G, Steenbergen AE, De Jong P. Spray drying of food products: 2. Prediction of insolubility index. Journal of Food Engineering. 1999b;**42**:73-77

[31] Büchi Labortechnik AG. Operation Manual – Mini Spray Dryer 190. Flawil; 1997-1998

[32] Beddow JK. Testing and Characterization of Powders and Fine Particles. London: Heyden; 1980. p. 195

[33] Omobuwajo TO, Busari OT, Osemwegie AA. Thermal agglomeration of chocolate drink powder. Journal of Food Engineering. 2000;**46**(2):73-81

[34] Hough G, Sánchez R. Descriptive analysis and external preference mapping of powdered chocolate milk. Food Quality and Preference. 1998;**9**(4):197-204

[35] Agibert SAC, Lannes SCS. Dark chocolate added with high oleic peanut oil microcapsule. Journal of the Science of Food and Agriculture. 2018;**98**:5591-5597. DOI: 10.1002/jsfa.9102

[36] Geranpour M, Emam-Djomeh Z, Asadi G. Investigating the effects of spray drying conditions on the microencapsulation efficiency of pumpkin seed oil. Journal of Food Processing & Preservation. 2019;**43**:13947

[37] Haas K, Obernberger J, Zehetner E, Kiesslich A, Volkert M, Jaeger H. Impact of powder particle structure on the oxidation stability and color of encapsulated crystalline and emulsified carotenoids in carrot concentrate powders. Journal of Food Engineering. 2019;**263**:398-408

[38] Jafari SM, Masoudi S, Akbar B. A Taguchi approach production of spray-dried whey powder enriched with nanoencapcapsulated vitamin D-3. Drying Technology. 2019;**37**(16):2059-2071

[39] Ulas B, Feyza E, Mehmet K. The effect of spray drying conditions on physicochemical properties of encapsulated propolis powder. Journal of Food Process Engineering. 2019;**42**:13024

[40] Labuza TP. Sorption phenomena in foods. Journal of Food Technology. 1968;**22**:263-271

[41] Labuza TP, Hyman CR. Moisture migration and control in multi-domain foods. Trends in Food Science and Technology. 1998;**9**:47-55

[42] Cornilon P, Salim LC. Characterization of water mobility and distribution in low- and intermediate-moisture food systems. Magnetic Resonance Imaging. 2000;**18**(3):335-341

[43] Bensebia O, Allia K. Analysis of adsorption-desorption moisture isotherms of rosemary leaves. Journal of Applied Research on Medicinal and Aromatic Plants. 2016;**3**:79-86

[44] Sormoli ME, Langrish TAG. Spray drying bioactive orange-peel extracts produced by Soxhlet extraction: Use of WPI, antioxidant activity and moisture sorption isotherms. LWT-Food Science and Technology. 2016;**72**:1-8

Chapter 4

Valorization of the Seeds (Almonds and Oil) of the Spontaneous Argan of Tindouf and the Other Experimental Domesticated Argan of Mostaganem in Algeria

Benaouf Zohra, Djorf Oussama, Jaradat Chawkat and Kechairi Reda

Abstract

The aim of the research was to determine the phytochemical and parameters of argan oil and almonds. We are interested in following the formation of volatile compounds in argan oil and also the determination of antioxidants; the purpose was mainly to identify and quantify the antioxidants to meet this objective, two samples of argan oil from the almonds of Tindouf taxa and Mostaganem taxa. The results show that the argan rich in phenolic compounds deserve to be exploited as much as nutritional and pharmaceutical supplements because of their antioxidant properties, which can surely contribute to the safeguarding of the argan tree.

Keywords: argan oil, almonds, volatile compounds, parameters, phenolics, antioxidant

1. Introduction

The argan tree, being a xerophile species, observed on the semiarid and arid climate has specific ecological characteristics and many interests (forest, forage, and fruit). Argan oil is essentially rich in unsaturated fatty acids and saturated fatty acids. As for the secondary metabolism, it contains polyphenols, tocopherol, sterol, and alcohol, and this explains its benefits in treating heart diseases and skin infections and in general its therapeutic uses and medication as a food supplement [1, 2]. Argan oil has a high level of oleic and linoleic acids and antioxidant compounds, which has an impact on cardiovascular disease [3]. Minor compounds of argan oil, such as sterols, may be involved in its cholesterol-lowering effect [4]. The antidiabetic effect of argan oil has been claimed for a long time in traditional medicine; however the mechanism of regulation of the level of glucose in the blood remains unknown [5]. The antihypertensive effect of argan oil and its mechanism of action have been studied by Berrougui et al. [6]. The purpose of this present work is the

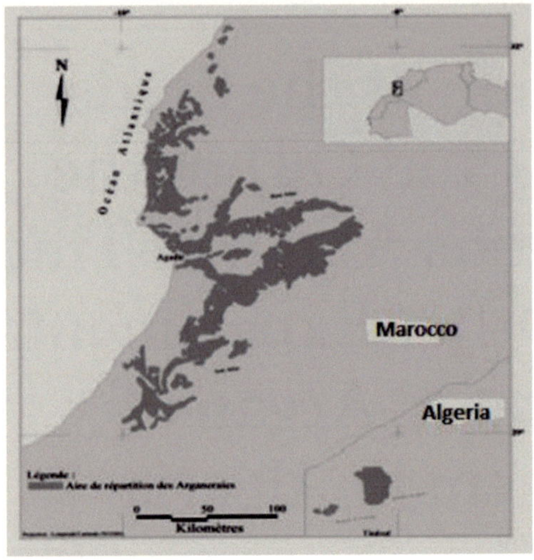

OR

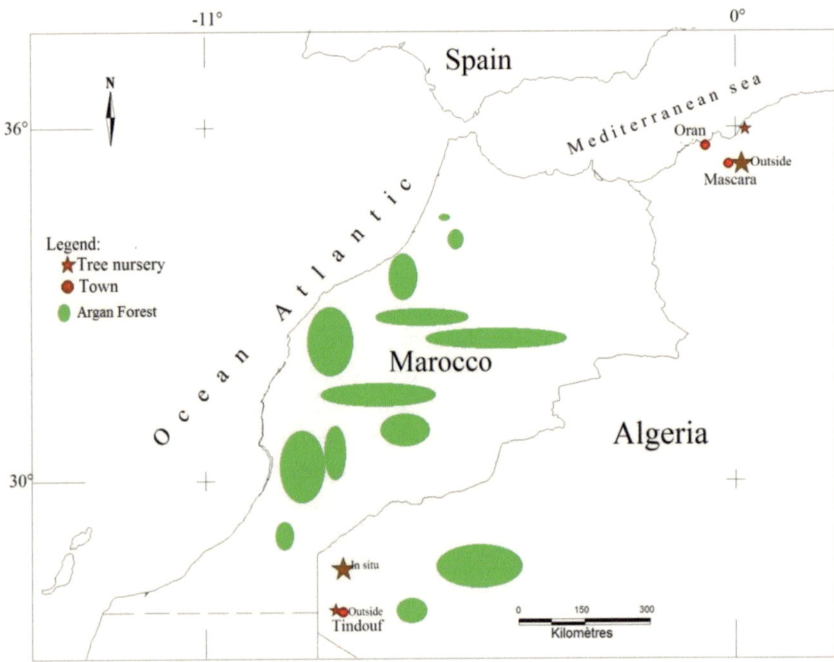

Figure 1.
Distribution of the argan tree in Tindouf and northwest Africa [7].

comparison between two provenances of argan tree, an endemic variety that grows in southwestern Tindouf located in Algerian Sahara and the other introduced to Mostaganem located in Mediterranean area (**Figure 1**).

2. Materials and methods

2.1 Plant material

A mature fruit of Algerian argan (*Argania spinosa*) was collected from Tindouf area located in Tindouf and Mostaganem (coastal region) in June 2016; almonds and the extracted oil were analyzed.

2.2 Argan oil preparation

The extraction was carried out by a Soxhlet apparatus, according to the standard technique [8]; the technique consists in using an organic solvent (hexane). 25 g of almond seeds powder are placed in a cartridge, and then the cartridge is closed by cotton and placed in the Soxhlet extractor. A flask is weighed empty and then filled with 200 ml of solvent. This flask is inserted into the extractor and placed in a sand bath set at a boiling point of the solvent. The extraction is carried out for 3 h and 6 h, then the solvent is removed by distillation, and the oil which remains in the flask is dried at a temperature of 105°C for a few minutes. The volatile compounds were extracted by the solid-phase microextraction (SPME) method; this technique does not require the use of solvents or complicated apparatus, and it is based essentially on the adsorption phenomenon based on a balance between the matrix and coating of the fiber. The identification and quantification of aromatic compounds were performed by gas chromatography-mass spectrometry (GC-MS) (**Figure 2**).

The chemical parameters were detected according to ISO standards [9, 10]. Total sugars were measured according to the method of Dubois et al. [11].

Equipment related to the results presented in **Table 1:** 50 mg of the almonds of each sample and put in a vial, distilled water is added until at 50 ml. Introduce 1 ml of the solution to be assayed into a tube of each sample and then 1 ml of the phenol solution (5%). The tubes are carefully shaken, and then 5 ml of concentrated

Figure 2.
Fruit, seed, and almond argan oil.

Parameters	Tindouf argan	Mostaganem argan
Relative density	0.83 ± 0.02	0.91 ± 0.03
Refractive index	1.4642 ± 0.08	1.4612 ± 0.04
Acid number	2.244 ± 0.01	2.524 ± 0.09
Index saponification	179.55 ± 0.8	185.60 ± 0.5
Ester index	177.306 ± 0.3	183.076 ± 0.5
pH	4.62 ± 0.07	4.43 ± 0.02
Humidity	2 ± 0.001%	4.33 ± 0.002%
Phosphatide	11.4 ± 0.06%	13.8 ± 0.02%
Extraction yield (3 h)	25.727 ± 0.08%	25.727 ± 0.02%
Extraction yield (6 h)	25.727 ± 0.07%	41.67 ± 0.04%
Total sugar	8.19 ± 0.04%	4.86 ± 0.06%
Fat	38.61 ± 0.3%	41.67 ± 0.5%
Ash	2.4 ± 0.03%	1.4 ± 0.01%
Nitrogen content	1.045 ± 0.001%	0.602 ± 0.00%
Protein	6.53 ± 0.04%	3.76 ± 0.005%

Table 1.
Chemical and physical parameters of argan oil of two taxa, Tindouf and Mostaganem argan.

sulfuric acid "H_2SO_4" are added using a graduated pipette. After standing for 30 min in the dark, the absorbance (OD) measurements are made at 490 nm in the case of hexoses. The calibration of the spectrophotometer (UV-vis spectrophotometer) is done with a blank solution containing 1 ml of distilled water, 1 ml of 5% phenol, and 5 ml of H_2SO_4.

The mineral material was determined by 5 g of the almonds of each region placed in the capsules and placed in the muffle furnace with a temperature of 900°C for 2 h and then metered in desiccators until it was cooled and finally weighed. The protein content is carried out in three stages: mineralization, distillation, and titration; in each flask 3 g of sample from each region are introduced, and 1.5 g of the catalyst is added with some glass bead, then 20 ml of sulfuric acid are poured in. A concentrated 50 ml of distilled water and 45 ml of sodium hydroxide solution (40%) are added for 3 min. The end of the apparatus is leveled in a tarpaulin containing 20 ml of boric acid (4%) which fixes the solution. The titration is carried out with a 0.1 N sulfuric acid solution in the presence of a colored indicator (methyl red) until a pink turn is obtained. For the determination of the fat, introduce 50 g of the sample from each region into the cartridge, place it in the Soxhlet, and weigh the empty flask and fill it with hexane (300 ml). After 6 h of extraction, determine the moisture by the loss of sample water (oil), take two capsules, put in each capsule 3 g of the oil, and placed in an oven at a temperature of 105°C for 2 h of drying. For the determination of the refractive index, calibrate the refractometer apparatus with distilled water. Then, one or two drops of each oil sample are placed on the prism, and the dark zone is moved in the middle for the separation cloth of the light and dark beach.

Determination of PH and acid number, 2 g of the sample and introduce it into the flask or flask. Add 5 ml of ethanol and some drops of phenolphthalein solution (or phenol red) as an indicator, and titrate the liquid with the potassium hydroxide solution contained in the burette to the color curve where the volume V is recorded. For the determination of saponification index, in a flask introduce 2 g of the sample, and add with a burette 25 ml of potassium hydroxide solution and fragments of pumice or porcelain. Fit the glass tube or refrigerant, and place the balloon on the boiling water bath. Allow to cool, disassemble the tube, and add 20 ml of water then 5 drops of phenolphthalein solution. The ester number is the number of milligrams of potassium hydroxide necessary for the neutralization of the acids released by the hydrolysis of the esters contained in 1 g of argan oil. Hydrolysis of the esters by heating in the presence of an ethanoic solution, determination of the excess of alkali by a standard solution of hydrochloric acid. For the determination of phosphatide content, introduce 25 g of oil and 200 ml of acetone in a flask, then leave the mixture at a temperature of 4°C for 2 hours, then filter the mixture on the paper previously weighed and dry this paper at a temperature of 100°C up to 150°C, and finally put in the desiccator and weigh.

2.2.1 SPME sampling conditions

Analysis was performed as described by Baccouri et al. [12]. Each oil sample was spiked with 4-methyl-2-pentanone (internal standard) to a final concentration of 6.7 μg/kg. Then 1.5 g was introduced into a 10 ml vial fitted with a silicone septum. The vial was immersed in a water bath at 40°C, and the oily solution is maintained under magnetic stirring. After 2 min, a divinylbenzene/carboxen/polydimethylsiloxane (DVB/CAR/PDMS) fiber (50/30 μm, 2 cm long from Supelco Ltd., Bellefonte, PA) was exposed to the sample headspace for 30 min [13] and immediately desorbed for 2 min at 260°C in the gas chromatograph in splitless condition. All the analyses were performed in triplicate.

2.2.2 GC-MS analysis

GC-MS analysis was performed with a Shimadzu GC-2010 gas chromatograph equipped with a Shimadzu QP-2010 Plus quadrupole mass spectrometer (Shimadzu Corporation, Kyoto, Japan) and a DB-WAXETR capillary column (30 m × 0.25 mm, 0.25 mm film thickness, J&W Scientific Inc., Folsom, CA, USA). Due to the high boiling point of the oily compounds, direct injection to GC-MS apparatus is impossible, and pre-preparation has to be done. We used increased temperatures. Detection was carried out by electron impact mass spectrometry in total ion current (TIC) mode, using ionization energy of 70 eV. The identification of volatile compounds was confirmed by the injection of pure standards. Compounds for which pure standards were not available were identified on the basis of mass spectra and retention indices available in the literature. The relative concentration (μg kg^{-1} of oil) of the identified compounds was calculated by relating the areas of the internal standard of each compound.

2.3 Spectrophotometric assays

2.3.1 Determination of total phenolic content (TPC)

The amount of total phenolics was assayed spectrophotometrically by means of the modified Folin-Ciocalteu method [14, 15]. Briefly, 2.5 ml of 10-fold diluted Folin-Ciocalteu reagent, 2 ml of 7.5% aqueous sodium carbonate solution, and 0.5 ml of phenolic extract were mixed well. After 15 min of heating at 45°C, the absorbance was measured at 765 nm with a UV-visible spectrophotometer (UV-1700 Pharmaspec, Shimadzu, Milan, Italy) [16].

2.3.2 Determination of DPPH radical scavenging activity (RSA)

The hydrogen-donating ability of the crude extract and radical scavenging activity (RSA) of argan fruit parts were investigated using the 2,2-diphenyl-1-picrylhydrazyl (DPPH•) radical scavenging assay (RSA) [17, 18]. All operations were done in the dark or dim light [19]. For control purpose, the absorbance of the DPPH• without samples was measured.

The inhibition percentage (IP) of the DPPH• by the extracts was calculated according the formula $IP = [(A0min - A60min)/A0min] \times 100$. With the percentage of remaining DPPH• being proportional to the antioxidant concentration in the extracts, the DPPH• scavenging activity was expressed as μM of Trolox equivalent (TE) per mg of sample.

2.3.3 Determination of total antioxidant activity (TAA)

The TAA in crude extracts was determined according to the Trolox equivalent antioxidant capacity (TEAC) assay following the original analytical procedure described by Re et al. [20] with slight modifications. ABTS radical cation (ABTS•+) was produced by reacting a 7 mM ABTS stock solution with 2.45 mM potassium persulfate (final concentration). For the study, the ABTS•+ stock solution was diluted with ethanol to an absorbance of 0.70 (±0.02) at 734 nm and equilibrated at 30°C. Sample solutions of 30 μL (or standard) were mixed with ABTS•+ solution 3 ml. Absorbance readings were taken at 30°C exactly 6 min after initial mixing. An appropriate solvent blank was obtained by mixing an absolute ethanol of 30 μL with ABTS•+ solution of 3 ml and monitored its absorbance at 6 min. All determinations were carried out in triplicate. The ABTS•+ scavenging effect (% Inhibition) was calculated by the equation % Inhibition = [(*A*734blank × *A*734sample)/*A*734blank] × 100 where *A*734blank and *A*734sample are the absorbances of ABTS•+ solution at 734 nm before and after the sample addition. Calibration was performed, as described previously, with Trolox stock solutions. Results were expressed as μM Trolox equivalent (TE) per mg of sample.

2.4 Statistical analysis

Significant differences among different oils were tested by the one-way analysis of variance and the Duncan test for mean comparison. Statistical analyses were performed using the software package Statistica version 7. Results were reported as mean ± standard deviation ($n = 3$). The analysis of variance (one-way ANOVA) was performed with SPSS software (version 12.0 for Windows, SPSS Inc., Chicago, Illinois). Duncan's test was applied to assess significant differences among the variables ($p < 0.05$), while Pearson correlation test was used to show their correlations.

3. Results and discussion

We noted a small difference in the extraction yield of almonds for both Tindouf and Mostaganem taxa, respectively (25.727–29.272%), after the 3 h duration, while they have a difference of 38.63–41.67% for the duration of 6 h. The percentages of the total sugar of Mostaganem taxa kernels (4.86%) are equal to almost half of the percentage of Tindouf taxa total sugars (8.19%). The sample of Tindouf taxa kernels contains a significant ash (2.4%) compared to the Mostaganem taxa sample (1.4%). The protein content of Tindouf almonds (6.35%) is high compared to the Mostaganem taxa (3.76%). As for the amount of fat, it is brought closer together between the two samples; the kernel gives a significant amount of (40%). Concerning the physicochemical characteristics of argan oil, according to our results we notice that argan oil is not miscible with ethanol for both samples. A difference in the moisture content between the argan oil of the Mostaganem taxa (4.33%) and the Tindouf taxa (2%), and the relative density, by comparison the Tindouf oil (0.83) is lower than Mostaganem (0.91). On the other hand, for the refractive index of argan oil, we record the same values 1.46) with an acid pH. Regarding the other indices, the acidity index of the oil of Mostaganem taxa is equal to 2.5245 and that of Tindouf is 2.2440; the saponification index of the oil of Mostaganem is 185.6 and that of Tindouf is 179.55 (**Table 2**).

The results show some volatile compounds, including compounds of lipid peroxidation, Strecker degradation, and Maillard reaction, responsible for the formation of pyrazines and autoxidation of fatty acids. This study could help to adjust the argan oil aroma and perhaps meet new types of consumers. For the phytochemical part, we have undertaken a study on the volatile composition of argan oil. Extraction of the volatile compounds was carried out by solid-phase microextraction (SPME), and their identification and quantification were performed by gas chromatography-mass spectrometry (GC-MS). Finally, we were interested in the elucidation and quantification of polyphenols. These secondary metabolites are of great importance because of their antioxidant properties. In total, 11 phenolic compounds were identified and quantified in the argan tree. This could be achieved through the coupling of liquid chromatography and electrospray negative ion mass spectrometry (LC-ESI-MS). Among the polyphenols cited are procyanidins B1 and B2, (+)-catechin, (−)-epigallocatechin gallate, (−)-epicatechin, isoquercitrin, hyperoside, rutin, phloridzin, myricetin, and quercitrin. The unroasted kernels and the shell are characterized by a diverse phenolic composition. The pulp is

Compound	Tindouf argan	Mostaganem argan	p
Acids			
Butanoic	1.2 ± 0.2d	1.5 ± 0.5d	**
Valeric	1.2 ± 0.3b	1.5 ± 0.2d	***
Hexanoic	5.2 ± 1.5b	4.9 ± 2.6b	**
Alcohols			
1-Butanol	1.6 ± 0.02	1.5 ± 0.43	ns
1-Pentanol	7.5 ± 0.4b	11.6 ± 2.4c	***
1-Hexanol	20.8 ± 1.5	20.5 ± 2.6	ns
2-Heptanol	0.7 ± 0.1a	1.1 ± 0.4a	***
2,3-Butanediol *d,l*	40.2 ± 4.3	34.3 ± 5.2	ns
Aldehydes			
Hexanal	26.1 ± 0.8b	28.9 ± 5.89	***
Benzaldehyde	1.2 ± 0.1bc	1.4 ± 0.9c	***
Esters, lactones			
Ethyl 2-methyl butanoate	0.1 ± 0.01ab	0.1 ± 0.03a	***
i-Amyl acetate	1.2 ± 0.03c	0.8 ± 0.4c	**
μ-Butyrolactone	2.2 ± 0.6c	3.6 ± 1.2c	**
Ketones			
2-Heptanone	5.2 ± 0.4b	4.8 ± 1.2c	***
Acetoin	5.4 ± 0.8ab	4.2 ± 1.4ab	***
2-Undecanone	4.2 ± 0.2ab	3.1 ± 0.6ab	***
Terpene			
Limonene	0.3 ± 0.4a	0.25 ± 0.6b	**
N-heterocycle μg/kg (it is not %)			
1-Methyl-*1H*-pyrrol	158.4 ± 22.1a μg/kg	147.1 ± 15.8b	**
2-Methyl pyrazine	121.3 ± 34.0ab	138.2 ± 36.5c	***
2,6-Dimethyl pyrazine	263.4 ± 42bc	273.1 ± 40.8d	**
2,3-Dimethyl pyrazine	6.3 ± 0.1ab	7.5 ± 0.9c	**
2-Ethyl-5-methyl pyrazine	24.9 ± 1.5a	23.4 ± 1.6c	***
2-Ethyl-6-methyl pyrazine	35.3 ± 4.2b	37.5 ± 3.2c	***
Trimethyl pyrazine	26.4 ± 1.2bc	49.8 ± 3.2d	***
2-Ethyl-3,5-dimethyl pyrazine	26.9 ± 0.5c	33.2 ± 0.9d	***
Furans			
2-Pentyl furan	23.1 ± 0.5ab	22.5 ± 5.8b	*
Furfurol	72.3 ± 1.0a	85.4 ± 4.4b	***
2-Furanmethanol	12.6 ± 0.7b	14.5 ± 2c	***

q, quantifier ion. Different letters in the same row at mean concentration values indicate significant differences ($p < 0.05$) as analyzed by Duncan test.
$^{*}p < 0.05$.
$^{**}p < 0.01$.
$^{***}p < 0.001$.

Table 2.
Quantified volatile compounds (μg/kg of oil ± SD) isolated in argan oil of two taxa.

quantitatively rich in total polyphenols (69.53 mg gallic acid equivalent/g). It showed a free radical scavenging activity, measured by DPPH. Important relative to other parts of the fruit (0.12 ± 0.004 μM Trolox equivalents/mg) and antioxidant activity (ABTS•+) (0.287 ± 0.05 μM equivalent/mg Trolox). Interestingly, the results obtained confirm that argan fruit polyphenols deserve to be exploited as much as nutritional and pharmaceutical supplements because of their antioxidant properties, which can surely contribute to the safeguarding of the argan tree. The aim of this work was to identify and quantify the phenolic compounds of argan fruit and by-products of argan oil extraction. Total phenolic content and antioxidant activity by DPPH and ABTS were evaluated. The LC-MS examination resulted in the detection of 10 compounds of which 8 were unambiguously identified. The identified compounds are classified into three groups: flavanols, flavonols, and dihydrochalcones. The results showed that six compounds were detected in the pulp: isoquercitrin and hyperoside are predominant (25.8 and 18.5 mg/100 g, respectively); they are followed by rutin (7.2 mg/100 g) and quercitrin (0.32 mg/100 g). Epicatechin and procyanidin B2 were also detected but could not be quantified. The phenolic compounds of the fruit shell of the argan tree have not been the subject of any prior work. The major phenolic compound isolated from the shell is (−)-epicatechin (0.45 mg/100 g), followed by isoquercitrin (0.32 mg/100 g). Rutin and phloridzin have the same level (0.18 mg/100 g), hyperoside and procyanidins B1 and B2 both 0.08 mg/100 g, myricetin 0.04 mg/100 g, and finally the quercitrin that was detected could not be quantified. As for kernels and meal, a major compound was detected; however this compound could not be identified by Tandem mass spectrometry (Mw = 423.5, Rt: 8.5 min).

4. Conclusion

The average oil density of Mostaganem taxa seems low compared to Tindouf oil. For the refractive index, a small difference is noted between the two taxa of *Argania spinosa* oil. For the acid index, the sample of the Mostaganem area seems to be richer in free fatty acids than in the Tindouf area. Same for the saponification index and the ester index, it is important in the argan oil of Mostaganem than Tindouf. For the phosphatide content, it seems high in both zones, but Mostaganem oil is richer. Also our results show that the phenolic fractions studied have remarkable antioxidant properties. Although the composition of the phenolic fraction of fruits can evolve over the years, they deserve a better valuation in the pharmacological, cosmetic, and agro-food fields because of their antioxidant properties.

Author details

Benaouf Zohra[1,2*], Djorf Oussama[3,4], Jaradat Chawkat[5] and Kechairi Reda[6]

1 Research Laboratory in Geo Environment and Spatial Development LGEDE, University of Mustapha Stambouli, Mascara, Algeria

2 Laboratory of Plant Ecology and Environment, Faculty of Biological Sciences, USTHB University, Algiers, Algeria

3 Laboratory of Biochemistry, Faculty of Chemistry, USTHB University, Algiers, Algeria

4 Laboratory of Ecology, Algeria

5 Research Laboratory on Biological, El Khalil, Palestine

6 Laboratory of Plant Ecology and Environment, Faculty of Biological Sciences, Tlemcen University, Tlemcen, Algeria

*Address all correspondence to: zohrabenaouf@gmail.com

References

[1] Hilali M, Charrouf Z, Soulhi AEA, Hachimi L, Guillaume D. Influence of origin and extraction method on argan oil physicochemical characteristics and composition. Journal of Agricultural and Food Chemistry. 2005;**53**:2081-2087

[2] Harhar H, Gharby S, Kartah B, El Monfalouti H, Guillaume D, Charrouf Z. Influence of argan kernel roasting-time on virgin argan oil composition and oxidative stability. Plant Food for Human Nutrition. 2011;**66**:163-168

[3] Derouiche A, Cherki M, Drissi A, Bamou Y, El Messal AM, Idrissi-Oudghiri A, et al. Nutritional intervention study with argan oil in man: Effects on lipids and apolipoproteins. Annals of Nutrition and Metabolism. 2005;**49**:196-201

[4] Khallouki F et al. Consumption of argan oil (Morocco) with its unique profile of fatty acids, tocopherols, squalene, sterols and phenolic compounds should confer valuable cancer chemopreventive effects. European Journal of Cancer Prevention. 2003;**12**:67-75

[5] Emonard H et al. Inhibition of gelatinase A by oleic acid. Annals of the New York Academy of Sciences. 1999;**878**:647-649

[6] Berrougui H, Alvarez de Sotomayor M, Perz-Guerrero C, Attaib A, Hmamouchi M, Marhuenda ER, et al. Argan (Argania spinosa) oil lowers blood pressure and improves endothelial dysfunction in spontaneously hypertensive rats. The British Journal of Nutrition. 2004;**92**:921-929

[7] Kechairi R, Abdoun F. Cartographic status of the argan tree *Argania spinosa* (L.) Skeels (Sapotaceae) in North-West Africa (Algeria and Western Sahara). International Journal of Environmental Studies. 2016;**73**:286-293

[8] Afnor. Compendium of French Standards on Fats, Oilseeds, Derived Products. 4th ed. Paris: French Association of Standardization; 1988

[9] ISO-6886. Animal and vegetable fats oils-determination of oxidative stability (accelerated oxidations test). 1989

[10] ISO 659. Oilseeds-Determination of oil content (reference method). 2009

[11] Dubois M et al. Colorimetric method for determination of suger and related substances. Analytical Chemistry. 1956:23-38

[12] Baccouri O, Bendini A, Cerretani L, Guerfel M, Baccouri B, Lercker G, et al. Comparative study on volatile compounds from Tunisian and Sicilian monovarietal virgin olive oils. Food Chemistry. 2008;**111**:322-328

[13] Naczk N, Shahidi F. Phenolics in cereals, fruits and vegetables: Occurrence, extraction and analysis. Journal of Pharmaceutical and Biomedical Analysis. 2006;**41**:1523-1542

[14] Singleton VL, Rossi JA. Colorimetry of total phenolics with phosphomolybdic phosphotungstic acid reagents. American Journal of Enology and Viticulture. 1965;**16**:144-158

[15] Singleton VL, Orthofer R, Lamuela-Raventos RM. Analysis of total phenols and other oxidation substrates and antioxidants by means of Folin-Ciocalteu reagent. Methods in Enzymology. 1999;**299**:152-178

[16] Pinelo M, Rubilar M, Jerez M, Sineiro J, Núñez MJ. Effect of solvent, temperature and solvent-to-solid ratio on the total phenolic content and

antiradicalary activity of extracts from different components of grape pomace. Journal of Agricultural and Food Chemistry. 2005;**53**:2111-2117

[17] Von Gadow A, Joubert E, Hansmann CF. Comparison of the antioxidant activity of aspalathin with that of other plant phenols of Rooibosed tea (*Aspalathus linearis*), α-tocopherol, BHT and BHA. Journal of Agricultural and Food Chemistry. 1997;**45**:632-638

[18] Schinella SM, Cienfuegos-Jovellanos E, Pasamar MA, Muguerza B, Ramón D, Ríos JL. Antioxidant properties of polyphenol-rich cocoa products industrially processed. Food Research International. 2010;**43**:1614-1623

[19] Sharma OP, Bhat TK. DPPH antioxidant assay revisited. Food Chemistry. 2009;**113**:1202-1205

[20] Re R, Pellegrini N, Proteggente A, Pannala A, Yang M, Rice-Evans C. Antioxidant activity applying an improved ABTS radical cation decolourization assay. Free Radical Biology and Medicine. 1999;**26**:1231-1237

Section 3

Quality of Raw Materials in Food Processing

Chapter 5

Microbiological Quality of Chicken Meat Fed with Olive Leaves (*Olea europaea L.*)

Cristiane Marangoni, Alexandre José Cichoski and Juliano Smanioto Barin

Abstract

This study evaluated the antimicrobial activity of olive leaves in vitro and meat chicken fed with percentages of 5 and 10 g of olive leaves for each kg feed. This is justified by the relevance of obtained safe products, with emphasis on the use of natural additives. The olive leaves presented antibacterial activity in all tested bacteria. For the bacteria *Yersinia enterocolitica*, *Escherichia coli*, and *Shigella*, the minimal inhibitory concentration varied from 0.6 to 1.5 mg/ml. The treatment with an addition of olive leaves showed better microbiological stability of the thighs and drumsticks of chickens than treatment without an addition of olive leaves. The use of 5 g/kg diet inhibited the growth of *Staphylococcus aureus* and *aerobic psychrotrophic* and aerobic mesophilic, while the use of 10 g/kg of diet inhibited the growth of *Enterococcus* spp., lactic bacteria, thermotolerant coliforms, *Pseudomonas*, *Clostridium perfringens,* and *Escherichia coli*.

Keywords: antimicrobial activity, olive leaves, natural additive, chicken meat

1. Introduction

In order to inhibit microbial growth of raw materials or cuts, it is often more effective than the direct addition of preservatives to add compounds into the diet of the growing animals [1]. The phenolic compounds occurring naturally in the plants have the ability to inhibit the growth of microorganisms, including bacteria, viruses, and fungi, maintaining the quality of meat for a long time.

In a study conducted by Bisignano et al. [2], the in vitro antimicrobial activity of oleuropein and hydroxytyrosol extracted from olive leaves was evaluated, and the efficiency for the pathogenic bacterium *Staphylococcus aureus* was identified. In another study Bisignano et al. [3] identified the antimicrobial components of olive leaves, and they discovered the effectiveness of long-chain aliphatic aldehydes against Gram-positive and Gram-negative bacteria.

In a study conducted by ERBAY and ICIER [4], the main compound found in olive leaves was oleuropein, being 24.54% in dry leaves. In a study realized by Paiva-Martins [5] with an objective of to assess the influence of OL supplementation at a lower level on feed digestibility and meat quality, the results indicated that olive leaves may be included in pig diets at 25 g/kg in order to improve the tocopherol content of meat without excessively compromising growth performance.

Upon investigating the in vitro activity of a commercial extract of olive leaf (*Olea europaea* L.) containing 4.4 mg/ml oleuropein, against a wide range of microorganisms, Sudjana et al. [6] determined that the compound has an inhibitory activity for *Helicobacter pylori*, *Campylobacter jejuni*, and *Staphylococcus aureus*. The in vitro activity of olive leaves was also studied by Markin et al. [7], who observed efficiency especially against *Klebsiella* and *Pseudomonas*.

Botsoglou et al. [8] evaluated the effect of the use of olive leaves in turkey's supplement diet in quantities of 5 and 10 g of leaves/kg diet in relation to microbiological quality of breast fillets that were stored at 4°C during 12 days. The turkey fillets that received olive leaves in the diet have had lower numbers of colonies of lactic acid bacteria, psychotropic, mesophilic, and enterobacteriaceae.

This study was designed to evaluate the effects of supplementation of the percentages of 5 and 10 g of olive leaves/kg of feed in the diet of broilers, on microbiological, of the meat the thighs and drumsticks stored at 4°C (±1°C) for 12 days.

2. Materials and methods

2.1 Extraction and chemical composition of olive leaves

Olive leaves (*Olea europaea* L.) of the variety Ascolana were collected between January and March 2012; drying in a tray dryer, at 45°C with air circulation for 72 hours, was realized. The leaves dried were milled in razor mill type Willey in 1 mm. The material was stored in paper and plastic packaging at 4°C until use.

The determination of total phenolics in olive leaves before followed the methodology described by Swain and Hills [9], which used as pattern the gallic acid, in concentrations of 50, 100, 150, 200, and 250 mg/l to build the calibration curve. Liquid chromatographic analysis of olive leaves.

We evaluated the oleuropein content present in olive leaves according to the method proposed by Guimarães et al. [10], and the chromatography conditions were based on Quirantes-pine et al. [11]. The separation of oleuropein was realized by using a HPLC Agilent 1260 Infinity (Agilent Technologies, Germany) liquid chromatography with a diode array detector (DAD).

2.2 Evaluation of different concentrations of olive leaves for antibacterial activity in vitro

The minimum inhibitory concentration (MIC) analysis for the in natura olive leaves (after harvest) and after drying for microorganisms was realized: *Escherichia coli* (ATCC8739), *Salmonella typhimurium* (ATCC14028), *Shigella dysenteriae* (NCTC7919), *Yersinia enterocolitica* (CDC175), *Clostridium perfringens* (NCTC8798), *Listeria monocytogenes* (ATCC19117), *Staphylococcus aureus* (ATCC29213), *Pseudomonas aeruginosa* (ATCC14502), and *Enterobacter aerogenes* (ATCC13048). The lyophilized bacteria were activated and replicated, and the suspension turbidity was standardized according to the nephelometric scale of McFarland in 0.5 which corresponds to the concentration of 1.5×10^8 CFU/ml (colony-forming units per milliliter).

The plant extract obtained from the olive leaves was evaluated according to microdilution in all concentrations: 20; 10; 5; 2.5; 1.25; 0.625; 0.312; and 0.156 mg/ml. The extract was put in plaques, and all the plaques were incubated in a greenhouse at 35°C for 24 hours and read with revealing. The read had objective show what concentrations the olive leaves had better effect on microorganisms.

2.3 Animals and diets

The chickens were created in the farm for 42 days and fed with the following diets: T1 (traditional diet without addition of olive leaves), T2 (diet with addition of 5 g of olive leaves for each kg feed), and T3 (diet with addition of 10 g of olive leaves for each kg of feed).

The broilers were slaughtered, and thighs and drumsticks, with skin and bone, were collected, stored in plastic bags of polyethylene without barrier, at 4°C (±1°C), for 12 days to monitor the growth microbiological.

2.4 Microbiological analysis

The poultry meat was analyzed microbiologically on days 0 (zero), 3, 6, 9, and 12 of storage.

Clostridium perfringens was performed with culture medium TSC and pour plate sowing depth and reading after 24 hours of incubation at 36°C (±1°C), according to the methodology described by IN 62, August 26, 2003, of the Ministry of Agriculture [12].

The analysis of fecal coliform, *Staphylococcus aureus*, aerobic mesophilic, *Escherichia coli*, *Enterococcus* spp., coliform bacteria, aerobic psychrotrophic, lactic acid, *Pseudomonas* spp., *Campylobacter* (*jejuni, coli,* and *lari*), *Salmonella*, and *Listeria monocytogenes* was performed according to the AOAC method [13].

Shigella, Streptococcus, Yersinia, and *Klebsiella* were determined in VITEK 2 [12].

2.5 Statistical analysis

All analyses took place in triplicate runs. Results were statistically analyzed by mean standard deviation, variance, and Tukey test at 95% significance, using the software Statistica 6.1 (Statsoft Inc., USA).

3. Results and discussion

3.1 Extraction and chemical composition of olive leaves

The average for the analysis of olive leaves was 4.65% moisture in dry basis, 4.69% of fixed mineral residue, 1.38% fat, 23.3% crude fiber, and 12.73 g/NT 6.25 × 100 g protein. This result is in accordance with that found by Botsoglou et al. [8]. The low percentage moisture of olive leaves ensures your quality, because it is not favorable to the development of fungi, molds, and yeasts.

The total phenolic content found in olive leaves, before and after the drying, was 12,275 and 9525 mg/g leaves, respectively. Similar values were found by Makris et al. [14] who reported 40.27 mg of gallic acid equivalents/g of dried olive leaves, and by Botsoglou et al. [8] who found phenol content of 26 mg of gallic acid equivalents/g of dried leaves.

The oleuropein tenor found in the olive leaves was 15.0 (±0.8) g/kg (CV de 5.1%, $n = 3$). This value was similar to the one found by Paiva-Martins et al. [15] which obtained 22.3 (±0.18) g/kg oleuropein in olive leaves.

3.2 In vitro antibacterial activity

Table 1 presented the values of inhibitory minimum concentration (MIC) in mg/ml from the olive leaf (*Olea europaea* L.) gross extract in natural and after drying.

Minimum inhibitory concentration—MIC (mg/ml)		
Microorganism	**In natural leaf extract**	**Dry leaf extract**
Salmonella typhimurium	20	>20
Staphylococcus aureus	20	>20
Pseudomonas aeruginosa	20	>20
Listeria monocytogenes	>20	>20
Enterobacter aerogenes	10	>20
Clostridium perfringens	5	>20
Shigella dysenteriae	1	0.156
Yersinia enterocolitica	0.625	0.156
Escherichia coli	0.625	0.078

Table 1.
TTest results for MIC determination for olive leaves (Olea europaea L.) extract.

All the bacteria tested presented sensibility for olive leaves, some with more intensity and others with less. For the bacteria *Yersinia enterocolitica*, *Escherichia coli*, and *Shigella*, both the olive extracts presented moderated action, with values between 0.6 and 1.5 mg/ml.

For the microorganisms *Salmonella typhimurium, Staphylococcus aureus, Pseudomonas aeruginosa, Listeria monocytogenes, Enterobacter aerogenes*, and *Clostridium perfringens*, a lower inhibitory action from olive extracts, with MIC above 1.5 mg/ml, was found. For the microorganisms that showed results >20, less capacity of inhibition was verified.

3.3 Effect of olive leaves on microorganisms in meat

The safety and quality of fresh broiler beef can be estimated by counting microorganism indicators aerobic mesophilic and psychrotrophic [16]. **Table 2** shows the number of colonies ($\log_{10}$ CFU/g) of aerobic mesophilic, lactic acid bacteria, aerobic psychrotrophic, and *Pseudomonas*.

Analyzing the growth of the microorganism aerobic mesophilic (**Table 2**), the treatments that received diet supplemented with olive leaves (T2 and T3) remained within the quality standards during the 12 days storage at 4°C reaching a maximum counting of 5.63 and 5.87 $\log_{10}$ CFU/g, respectively, while control treatment showed a counting of 6.07 $\log_{10}$ CFU/g from the third day of storage. The aerobic mesophilic counting of 10^7 CFU/g or 7 $\log_{10}$ CFU/g is considered an indicator for the end of shelf life of cooled broiler meat [17]. Some studies that are more precise indicate outside the ideal sanitary conditions broilers with a counting of mesophilic 10^6 CFU/g [18]. According to these parameters, the counting between 10^6 and 10^7 CFU/g was considered a limit to end the shelf life, and it can be said that the diet with added olive leaves of broilers provided an increase in the shelf life of meat compared with the control treatment.

The number of colonies of lactic acid bacteria, in the treatments with olive leaves (T2 and T3) in all analyzed days, was lower than that of the control treatment (T1) and differed significantly ($P < 0.05$) among themselves. These results show that olive leaves present an inhibitory effect on the growth of lactic acid bacteria (**Table 2**). Between treatments with olive leaves, T3 had throughout the period fewer colonies of lactic acid bacteria than T2, and this result was significantly different ($P < 0.05$). Botsoglou et al. [8] added 5 and 10 g olive leaves/kg in the

		Analysis			
		Aerobic mesophilic bacteria (log CFU/g)	**Lactic acid bacteria (log CFU/g)**	**Psychrotrophic bacteria (log CFU/g)**	***Pseudomonas* spp. (log CFU/g)**
Storage time (days)					
0	T1	7.30E + 05[a]	1.00E + 02[a]	1.10E + 03[a]	3.20E + 03[a]
		(±0.0002)	(±0.0002)	(±0.0002)	(±0.0002)
	T2	3.90E + 03[b]	4.40E + 01[b]	7.80E + 01[c]	7.01E + 02[b]
		(±0.0002)	(±0.0002)	(±0.0002)	(±0.0002)
	T3	1.30E + 03[c]	3.20E + 01[c]	6.20E + 02[b]	8.00E + 01[c]
		(±0.0002)	(±0.0002)	(±0.0002)	(±0.0002)
3	T1	1.20E + 06[a]	5.20E + 03[a]	2.30E + 04[a]	1.30E + 04[a]
		(±0.0002)	(±0.0002)	(±0.0002)	(±0.0002)
	T2	5.60E + 04[b]	7.32E + 02[b]	1.97E + 02[c]	3.80E + 03[b]
		(±0.0002)	(±0.0002)	(±0.0002)	(±0.0002)
	T3	2.90E + 04[c]	1.80E + 02[c]	2.10E + 03[b]	2.10E + 03[c]
		(±0.0002)	(±0.0002)	(±0.0002)	(±0.0002)
6	T1	2.50E + 06[a]	4.20E + 05[a]	2.81E + 04[a]	5.30E + 04[a]
		(±0.0002)	(±0.0002)	(±0.0002)	(±0.0002)
	T2	9.70E + 04[c]	3.50E + 03[b]	2.10E + 03[c]	4.70E + 03[b]
		(±0.0002)	(±0.0002)	(±0.0002)	(±0.0002)
	T3	6.60E + 05[b]	3.10E + 03[c]	1.71E + 04[b]	3.10E + 03[c]
		(±0.0002)	(±0.0002)	(±0.0002)	(±0.0002)
9	T1	1.91E + 07[a]	1.70E + 06[a]	1.30E + 05[a]	1.80E + 05[a]
		(±0.0002)	(±0.0002)	(±0.0002)	(±0.0002)
	T2	2.30E + 05[c]	2.30E + 04[b]	6.72E + 03[c]	2.30E + 04[b]
		(±0.0002)	(±0.0002)	(±0.0002)	(±0.0002)
	T3	6.81E + 05[b]	5.80E + 03[c]	2.40E + 04[b]	5.80E + 03[c]
		(±0.0002)	(±0.0002)	(±0.0002)	(±0.0002)
12	T1	4.51E + 07[a]	3.50E + 06[a]	3.80E + 05[a]	4.90E + 06[a]
		(±0.0002)	(±0.0002)	(±0.0002)	(±0.0002)
	T2	4.30E + 05[b]	3.10E + 04[b]	7.42E + 03[c]	3.80E + 04[b]
		(±0.0002)	(±0.0002)	(±0.0002)	(±0.0002)
	T3	7.51E + 05[b]	6.00E + 03[c]	3.60E + 04[b]	6.20E + 03[c]
		(±0.0002)	(±0.0002)	(±0.0002)	(±0.0002)

T1 (control diet), T2 (diet supplemented with 5 g olive leaves/kg feed), and T3 (diet supplemented with 10 g olive leaves/kg feed). a, b, and c are scanned horizontally between T1, T2, and T3 intervals for analysis. Different letters show significant difference ($P < 0.05$) by Tukey test.

Table 2.
Number of colonies (log_{10} CFU/g) of aerobic mesophilic bacteria, lactic acid bacteria, psychrotrophic bacteria, and Pseudomonas spp. in thighs and drumsticks of broiler storage at 4°C for 12 days.

turkey feed and evaluated microbial growth in breast fillets, which were stored at 4°C for 12 days. On the twelfth day of storage, the number of lactic acid bacteria in the control treatment was 6.5 log_{10} CFU/g, and treatments with 10 and 5 g olive leaves were 4 and 5 log_{10} CFU/g, respectively. In this study, the number of lactic acid bacteria in meat on day 12 was 3.77 and 4.49 log_{10} CFU/g for treatments T3 and T2, respectively, and these values were lower than those found by Botsoglou et al. [8] in the same storage time and the same temperature.

Although the counting of aerobic psychrotrophic microorganisms indicates the degree of deterioration of refrigerated foods, the Brazilian legislation establishes no standard for these microorganisms. However, the International Commission on Microbiological Specifications for Foods [17] establishes 10^6–10^7 CFU/g as a standard. Considering these microbiological standards, the chicken meat in this study was fit for consumption during the 12 days at 4°C (**Table 2**). Throughout the storage period, T2 had the lowest number of colonies, with a significant difference ($P < 0.05$) of T3 and T1. The number of colonies of T3 was lower than that found in T1 also with significant difference ($P < 0.05$), indicating that olive leaves had significant effects on the growth of the counting of aerobic psychrotrophic microorganisms. In breast turkey fillets that received olive leaves in the diet, Botsoglou et al. [8] found the number of colonies of aerobic psychrotrophic on the twelfth day of storage of 4.6 and 5.7 $\log_{10}$ CFU/g for tests that received 10 and 5 g of olive leaves/kg diet. This counting were largest found that in this study, 4.55 to T3 and 3.87 $\log_{10}$ CFU/g to T2 (**Table 2**).

The analysis of total aerobes and *Pseudomonas* spp. are good indicators of spoilage of poultry meat [19]. The counting of *Pseudomonas* spp. is defined by several authors as indicating the end of useful life values when they reach 6–7 $\log_{10}$ CFU/g [20]. Considering these values and the results shown in **Table 2**, the three treatments were acceptable for consumption during the 12 days. The treatments T2 and T3 showed that a number of colonies of *Pseudomonas* spp. were significantly lower ($P < 0.05$) than that found in T1, on all days of storage. There were significant differences ($P < 0.05$) between treatments T2 and T3 throughout the study period, indicating that supplementation with olive leaves 5 g/kg feed showed better inhibitory capacity for this microorganism.

Table 3 shows the results of the microbiological analysis of total coliforms, *Enterococcus* spp., *Staphylococcus aureus*, thermotolerant coliforms, *Clostridium perfringens*, and *Escherichia coli*.

The Brazilian legislation does not establish microbiological parameters of coliforms. The treatments were subjected to this analysis to know the microbial load and so evaluate the sanitary conditions of the broiler meat of the three treatments, since these parameters reflect the quality of the raw material. The results vary between treatments (**Table 3**), where T2 and T3 had lower levels of total coliforms than T1 during the 12 days of storage.

The *Escherichia coli* presence in foods indicates microbial contamination of fecal origin [21]. The *Escherichia coli* (**Table 3**) started with similar values among the three treatments 3.07, 2.79, and 2.73 $\log_{10}$ CFU/g for T1, T2, and T3, respectively. After 9 and 12 days of monitoring, T2 and T3 had significant reductions ($P < 0.05$) compared to T1, demonstrating that the use of olive leaves at both concentrations had better effect inhibitory which T1. The use of 10 g/kg olive leaves in diet had greater inhibitory effect than the use of 5 g/kg for this microorganism.

The use of olive leaves in the amount of 10 g/kg feed showed a better inhibitory effect than the use of 5 g/kg feed for *Clostridium perfringens*, and both showed an inhibitory effect significantly ($P < 0.05$) better than T1.

According the Resolution no 12/2001, National Agency for Sanitary Surveillance [22], cuts of broiler cooled or frozen can have a tolerance limit for counting coliforms 45°C/g of 10^4 or 4 $\log_{10}$ CFU/g. According to this tolerance, T1 would be inappropriate for marketing and consumption because its initial counts were 4.5 $\log_{10}$ CFU/g, while the treatments with olive leaves have had initial counts of 1.93 and 2.36 for T3 and T2, respectively, and had presented counts within the tolerance limit of the legislation until the sixth day of storage at 4°C. The analysis

		Analysis					
		Total coliforms (log CFU/g)	***Enterococcus* spp. (log CFU/g)**	***Staphylococcus aureus* (log CFU/g)**	**Thermotolerant coliforms (log CFU/g)**	***Clostridium perfringens* (log CFU/g)**	***Escherichia coli* (log CFU/g)**
Storage time (days)							
0	T1	$4.40E + 02^{a}$	$3.31E + 03^{a}$	$3.80E + 01^{a}$	$3.20E + 04^{a}$	$5.00E + 00^{a}$	$1.20E + 03^{a}$
		(±0.0002)	(±0.0002)	(±0.0002)	(±0.0002)	(±0.0002)	(±0.0002)
	T2	$1.12E + 02^{b}$	$1.80E + 02^{b}$	$1.00E + 00^{c}$	$2.31E + 02^{b}$	$3.00E + 00^{a}$	$6.30E + 02^{b}$
		(±0.0002)	(±0.0002)	(±0.0002)	(±0.0002)	(±0.0002)	(±0.0002)
	T3	$8.60E + 01^{c}$	$1.10E + 02^{b}$	$2.30E + 01^{b}$	$8.70E + 01^{b}$	$2.00E + 00^{a}$	$5.41E + 02^{b}$
		(±0.0002)	(±0.0002)	(±0.0002)	(±0.0002)	(±0.0002)	(±0.0002)
3	T1	$6.10E + 03^{a}$	$4.20E + 04^{a}$	$4.21E + 02^{a}$	$5.50E + 05^{a}$	$1.07E + 01^{a}$	$3.70E + 04^{a}$
		(±0.0002)	(±0.0002)	(±0.0002)	(±0.0002)	(±0.0002)	(±0.0002)
	T2	$2.81E + 02^{b}$	$2.30E + 03^{b}$	$1.00E + 00^{c}$	$9.31E + 02^{b}$	$1.00E + 01^{a}$	$9.30E + 03^{b}$
		(±0.0002)	(±0.0002)	(±0.0002)	(±0.0002)	(±0.0002)	(±0.0002)
	T3	$1.52E + 02^{c}$	$3.41E + 02^{c}$	$4.83E + 01^{b}$	$9.51E + 02^{b}$	$1.00E + 01^{a}$	$7.30E + 03^{b}$
		(±0.0002)	(±0.0002)	(±0.0002)	(±0.0002)	(±0.0002)	(±0.0002)
6	T1	$9.71E + 03^{a}$	$1.60E + 05^{a}$	$7.80E + 03^{a}$	$7.40E + 05^{a}$	$1.20E + 04^{a}$	$7.40E + 05^{a}$
		(±0.0002)	(±0.0002)	(±0.0002)	(±0.0002)	(±0.0002)	(±0.0002)
	T2	$4.80E + 02^{b}$	$4.10E + 03^{b}$	$2.10E + 01^{c}$	$1.40E + 03^{c}$	$8.40E + 02^{b}$	$1.60E + 04^{b}$
		(±0.0002)	(±0.0002)	(±0.0002)	(±0.0002)	(±0.0002)	(±0.0002)
	T3	$1.50E + 02^{c}$	$1.20E + 03^{c}$	$2.83E + 02^{b}$	$3.80E + 03^{b}$	$2.30E + 02^{c}$	$1.10E + 04^{b}$
		(±0.0002)	(±0.0002)	(±0.0002)	(±0.0002)	(±0.0002)	(±0.0002)

		Analysis					
		Total coliforms (log CFU/g)	*Enterococcus* spp. (log CFU/g)	*Staphylococcus aureus* (log CFU/g)	Thermotolerant coliforms (log CFU/g)	*Clostridium perfringens* (log CFU/g)	*Escherichia coli* (log CFU/g)
9	T1	5.50E + 04^{a}	6.30E + 05^{a}	1.90E + 04^{a}	1.50E + 06^{a}	6.21E + 04^{a}	1.30E + 06^{a}
		(±0.0002)	(±0.0002)	(±0.0002)	(±0.0002)	(±0.0002)	(±0.0002)
	T2	3.10E + 03^{b}	4.20E + 04^{b}	5.01E + 02^{c}	7.50E + 04^{b}	3.20E + 03^{b}	1.90E + 05^{b}
		(±0.0002)	(±0.0002)	(±0.0002)	(±0.0002)	(±0.0002)	(±0.0002)
	T3	2.80E + 03^{c}	3.10E + 03^{c}	3.90E + 03^{b}	2.92E + 04^{c}	6.12E + 02^{c}	4.10E + 03^{c}
		(±0.0002)	(±0.0002)	(±0.0002)	(±0.0002)	(±0.0002)	(±0.0002)
12	T1	7.40E + 04^{a}	8.80E + 05^{a}	5.60E + 04^{a}	5.30E + 06^{a}	8.41E + 04^{a}	5.20E + 06^{a}
		(±0.0002)	(±0.0002)	(±0.0002)	(±0.0002)	(±0.0002)	(±0.0002)
	T2	4.10E + 03^{b}	5.20E + 04^{b}	7.01E + 02^{c}	9.60E + 04^{b}	5.40E + 03^{b}	2.20E + 05^{b}
		(±0.0002)	(±0.0002)	(±0.0002)	(±0.0002)	(±0.0002)	(±0.0002)
	T3	3.20E + 03^{c}	4.20E + 03^{c}	4.40E + 03^{b}	3.12E + 04^{c}	7.22E + 02^{c}	5.10E + 03^{c}
		(±0.0002)	(±0.0002)	(±0.0002)	(±0.0002)	(±0.0002)	(±0.0002)

T1 (control diet), T2 (diet supplemented with 5 g olive leaves/kg feed), and T3 (diet supplemented with 10 g olive leaves/kg feed). a, b, and c are scanned horizontally between T1, T2, and T3 intervals for analysis. Different letters show significant difference ($P < 0.05$) by Tukey test.

Table 3.
Number of colonies (log_{10} CFU/g) of total coliforms, Enterococcus spp., Clostridium perfringens, Staphylococcus aureus, thermotolerant coliforms, and Escherichia coli in thighs and drumsticks of broiler storage at 4°C for 12 days.

of fecal coliform indicated that T2 and T3 have had significant inhibitory effect ($P < 0.05$) compared to T1, demonstrating that both concentrations of olive leaves are inhibitory.

The research of *Enterococcus* spp. is not mandated by legislation, and few studies investigated these microorganisms. The counting showed that T3 had the best inhibitory effect than T2 and T1, and in the twelfth day, values of 3.61, 4.71 and 5.94 $\log_{10}$ CFU/g to T3, T2, and T1, respectively, were found. The broiler meat that received an addition of olive leaves (T2 and T3) showed significant reductions ($P < 0.05$) in the counts of *Enterococcus* spp. from the third to twelfth day of storage, demonstrating efficient reduction of this microorganism regarding the treatment of broilers that received a traditional diet (T1).

The current legislation in Brazil does not set a standard for *Staphylococcus aureus* in broiler meat; however, there are reports that are required between 10^5 and 10^6 CFU/g of *Staphylococcus aureus* per gram of food so that the toxin is formed at levels that can cause intoxication [21]. Considering this pattern, and analyzing the growth (**Table 3**), one can say that during the 12 days of storage at 4°C, the broilers meat of the treatments broilers fed the diet supplemented with the olive leaves have had significantly lower count ($P < 0.05$) than the T1, which is indicative of safer conditions. The maximum counts found for the broiler thighs and drumsticks were 2.84, 3.64, and 4.74 $\log_{10}$ CFU/g for T2, T3, and T1, respectively. Inhibition of *Staphylococcus aureus* was significantly lower ($P < 0.05$) in the treatment which received the addition of olive leaves in 5 g/kg of diet than T1 and T3. At the twelfth-day follow-up, the difference between T1 and T3 was 1.1 $\log_{10}$ CFU/g, demonstrating the inhibitory effect of olive leaves for this microorganism.

The federal legislation provides the absence of *Salmonella* in 25 g for poultry meat cooled. In the broiler thighs and drumsticks analyzed, the *Salmonella* wasn't present. This result confirms the microbial quality of the products, since the absence of *Salmonella* in samples attests to the hygienic and sanitary conditions.

The samples of broiler meat analysis of the three treatments showed absence of *Listeria monocytogenes*, proving the safety of the product for listeriosis.

The research for *Campylobacter* (*coli*, *jejuni*, and *lari*), *Shigella*, *Klebsiella*, *Yersinia*, and Streptococcus indicated the absence of these microorganisms for the three treatments during the study period.

4. Conclusions

The use of 5 g/kg olive leaves reduced the growth of *Staphylococcus aureus* and psychrotrophic total aerobic count, while 10 g/kg of diet reduced the growth of count of Enterobacteriaceae, lactic acid bacteria, total coliforms, *Pseudomonas* spp., *Clostridium perfringens*, and *Escherichia coli*. The samples of broiler meat analysis of the three treatments showed the absence of *Listeria monocytogenes.*

Author details

Cristiane Marangoni*, Alexandre José Cichoski and Juliano Smanioto Barin
Department of Science and Food Technology, Federal University of Santa Maria, Santa Maria, RS, Brazil

*Address all correspondence to: eng.cristiane@gmail.com

References

[1] Govaris A, Botsoglou N, Papageorgiou G, Botsoglou E, Amvrosiadis I. Dietary versus post-mortem use of oregano oil and/or a-tocopherol in turkeys to inhibit development of lipid oxidation in meat during refrigerated storage. International Journal of Food Sciences and Nutrition. 2004;**55**:115-123

[2] Bisignano G, Tomaino A, Locascio R, Crisafi G, Uccella N, Saija A. On the in vitro activity of oleuropein and hydroxytyrosol. The Journal of Pharmacy and Pharmacology. 1999;**51**:971-974

[3] Bisignano G, Lagana MG, Trombetta D, Arena S, Nostro A, Uccella N. In vitro antibacterial activity of some aliphatic aldehydes from *Olea europea* L. FEMS Microbiology Letters. 2001;**198**:9-13

[4] Erbay Z, Icier F. The importance and potential uses of olive leaves. Food Reviews International. 2010;**26**:319-334

[5] Paiva-Martins F, Ribeirinha T, Silva A, Gonçalves R, Pinheiro V, Mourão JL, et al. Effects of the dietary incorporation of olive leaves on growth performance, digestibility, blood parameters and meat quality of growing pigs. Journal of the Science of Food and Agriculture. 2014;**94**:14

[6] Sudjana AN, D'Orazio C, Ryan V, Rasool N, NG J, Islam N, et al. Antimicrobial activity of commercial Olea europaea (olive) leaf extract. International Journal of Antimicrobial Agents. 2009;**33**:461-463

[7] Markin D, Duek L, Berdicevsky I. In vitro antimicrobial activity of olive leaves. Mycoses. 2003;**46**:132-136

[8] Botsoglou E, Govaris A, Christaki E, Botsoglou N. Effect of dietary olive leaves and/or a-tocopheryl acetate supplementation on microbial growth and lipid oxidation of Turkey breast fillets during refrigerated storage. Food Chemistry. 2010;**121**:17-22

[9] Swain T, Hills WE. The phenolic constituents of Punnus domestica. I—quantitative analysis of phenolic constituents. Journal of the Science of Food and Agriculture. 1959;**19**:63-68

[10] Guimarães R, Barros L, Dueñas M, Calhela RC, Carvalho AM, Santos-Buelga C, et al. Nutrients, phytochemicals and bioactivity of wild Roman chamomile: A comparison between the herb and its preparations. Food Chemistry. 2013;**136**:718-725

[11] Quirantes-Piné R, Lozano-Sánchez J, Herrero M, Ibáñez E, Segura-Carretero A, Fernández-Gutiérrez A. HPLC-ESI-QTOF-MS as a powerful analytical tool for characterising phenolic compounds in olive-leaf extracts. Phytochemical Analysis. 2013;**24**:213-223

[12] Brazil. Ministry of Agriculture, Livestock and Food Supply, Secretariat of Agricultural and Livestock Defense, Normative Instruction No. 62, August 26, 2003. Manual of Official Analytical Methods for Microbiological Analysis for Control of Animal and Water Products. Official Diary of the Union Ministry of Agriculture, Livestock and Food Supply, Secretariat of Agricultural and Livestock Defense, Normative Instruction No. 62, August 26, 2003. Manual of Official Analytical Methods for Microbiological Analysis for Control of Animal and Water Products. Official Diary of the Union, Brasília, DF, 18 set. Seção 1, 2003; p. 14, Capítulo II; Disponível em: http://www.a3q.com.br/dmdocuments/Instru_Normativa_62.pdf. [Acesso em: 28 May 2015]

[13] AOAC. Association of Official Analytical Chemists. Official Methods

of Analysis of AOAC International. 18th ed. Gaithersburg: AOAC International; 2003

[14] Makris DP, Boskou G, Andrikopoulos NK. Polyphenolic content and in vitro antioxidant characteristics of wine industry and other agri-food solid waste extracts. Journal of Food Composition and Analysis. 2007;**20**:125-132

[15] Paiva-Martins F, Barbosa S, Pinheiro V, Mourão JL, Outor-Monteiro D. The effect of olive leaves supplementation on the feed digestibility, growth performances of pigs and quality of pork meat. Meat Science. 2009;**82**:438-443

[16] Jay JM. Microbiologia de Alimentos. 6th ed. Porto Alegre: Artmed; 2005

[17] ICMSF. International commission on microbiological specifications for foods. Microorganisms in Foods. Sampling for Microbiological Analysis: Principles and Scientific Applications. Toronto: University of Toronto; 2: 1986; pp. 181-186

[18] Ritter R, Bergmann GP. Eficácia do sistema de pré-resfriamento de frangos em tanques, sobre a redução da contaminação bacteriana de carcaças. Higiene alimentar. 2003;**17**:97-104

[19] Dominguez SA, Schaffner DW. Development and validation of a mathematical model to describe the growth of pseudomonas spp. in raw poultry stored under aerobic conditions. International Journal of Food Microbiology. 2007;**120**:287-295

[20] Mehyar GF, Blank G, Han JH, Hydamaka A, Holley RA. Effectiveness of trisodium phosphate, lactic acid and commercial antimicrobials against pathogenic bacteria on chicken skin. Food Protection Trends. 2005;**25**:351-362

[21] Franco BDGM, Landgraf M. Microbiologia dos Alimentos. São Paulo: Atheneu; 2005. p. 182

[22] Brazil. Ministry of Agriculture, Livestock and Food Supply, Secretariat of Agricultural and Livestock Defense, Normative Instruction No. 12 of April 10, 2001. Mercosur technical regulation analytical methodologies, ingestion of time loss and loss of production of veterinary foods in foods source. Animal. Official Journal of the Union, Brasília, DF, 12 abr. Seção 1. 2001

Chapter 6

Autochthonous Breeds of Republic of Serbia and Valuation in Food Industry: Opportunities and Challenges

Čedomir Radović, Milica Petrović, Marija Gogić, Dragan Radojković, Vladimir Živković, Nenad Stoiljković and Radomir Savić

Abstract

Climate change and the emergence of new animal diseases emphasize the need to maintain and conserve plant and animal resources because of their adaptive capacity. For hundreds of millions of poor rural households, livestock production and crop farming are a key resource for life. Livestock production has great contribution and is vitally important for the life and safety of crop production. Conservation of animal genetic resources is not easy and simple, but it is of concern that in the past some animal genetic resources have been lost before their characterization and their genetic potential has not been studied. It is known that with the loss of a single breed or strain, the genetic diversity contained within also disappears. That is why it is necessary to continuously work on the conservation of animal genetic resources using various methods of conservation. The preservation and improvement of livestock production and animal genetic diversity, the preservation and the development of locally adapted (autochthonous) new breeds, as well as the preservation of genetically diverse populations provide society with a greater range of options to meet future challenges and further develop agriculture.

Keywords: animal genetic resources, phenotype, Mangalitsa, Moravka, Resavka

1. Introduction

Pig farming in the Republic of Serbia has been of great importance since ancient times and represents an important branch of agricultural production. The importance of pig breeding in livestock production and the overall economy of Serbia comes from its economic and biological importance. Opportunities and conditions for growing and improving pig production are very favorable. Agricultural biodiversity is a product of thousands of years of activity in which people are trying to meet their needs in a wide range of social, climatic, and ecological conditions. Well-adapted and developed livestock breeding is an essential element of the agricultural production system; it is especially important in difficult conditions when the plant production in the agroecosystem cannot maintain and increase its productivity and adapt to changing circumstances and is crucial for the food safety of the population. The preservation and improvement of livestock production and animal genetic diversity, the preservation of and the development of locally adapted (autochthonous) new breeds, as well as the preservation of genetically diverse populations provide society with a greater range of options to meet future challenges and further develop agriculture. Responsible management of agricultural biodiversity in the world is becoming an increasing challenge for the international community, especially in the livestock sector, as it goes through dramatic changes by intensifying production, trying to respond to growing demands for increased production of meat, milk, and eggs. A wide range of many different animal genetic resources that correspond to and meet different human needs and desires is crucial for our adaptation and development of the agricultural production system. Climate change and the emergence of new animal diseases emphasize the need to maintain and conserve plant and animal resources because of their adaptive capacity. For hundreds of millions of poor rural households, livestock production and crop farming are a key resource for life. Livestock production has great contribution and is vitally important for the life and safety of crop production. Conservation of animal genetic resources is not easy and simple, but it is of concern that in the past some animal genetic resources have been lost before their characterization and their genetic potential has not been studied. It is known that with the loss of a single breed or strain, the genetic diversity contained within also disappears. That is why it is necessary to continuously work on the conservation of animal genetic resources using various methods of conservation.

In recent years, interest in autochthonous breeds has increased, not only for the purpose of preserving their genes but also for obtaining raw materials for the production of traditional dry meat products (Kulen sausage, bacon, pork skin crackling, etc.). This would allow the development of rural areas and small family farms.

2. The emergence of pig breeds: Mangalitsa (ML), Moravka (M), and Resavka

2.1 Mangalitsa

The first description for *Mangalitsa* found in scientific literature is from [1] 1886 in the journal *Težak*. *Mangalitsa* is improved Šumadinka breed, that is, Šumadinka breed which became more productive in improved conditions of growing and, above all, nutrition, care, and housing [2]. However, [3] states that the Mangalitsa was formed by the improvement of Šumadinka but also crossing with the Bakonian pig cultivated in Hungary. Mangalitsa was formed by crossing of the extinct Hungarian and Mediterranean breeds of pigs [4].

Genetic links between Hungarian Mangalitsa that are farmed in different geographic locations have been studied by 10 microsatellite markers. Estimated distances (Da, Ds, Fst) were the lowest between Swallow and White Mangalitsa, while Red Mangalitsa showed the highest genetic distance from the previous two breeds [5]. Based on the structure of DNA strains of the Mangalitsa pig, researchers have found that strains should be considered as individual breeds (Swallow Belly, White and Red Mangalitsa). Mangalitsa is a late breed of pigs, a fatty type with more strains within the breed (Swallow Belly, White and Red Mangalitsa). The occurrence of inbreeding is one of the main causes of poor production performance of this breed and has a negative impact on the condition of the population. There is a tendency to overcome these problems, which requires more systematic breeding and selection work (**Figure 1**).

Figure 1.
Swallow Belly Mangalitsa (Foto Č. Radović).

Figure 2.
Moravka (Foto Č. Radović).

2.2 Moravka

Moravka is a domestic pig breed cultivated mainly in the region of Svilajnac from which it spread throughout Serbia. It is the product of an unplanned crossing of Šumadinka and Berkshire pig breeds. Many years of random crossing and non-systematic selection have led to one stabilized type that became a breed. Random unplanned cultivation (unplanned use in reproduction and taking no consideration of relations) until the Second World War has led to further weakening of its body constitution and production. It imposed the adoption of certain measures from 1953 to 1958, such as the import of the Cornwall and its use for the improvement of Moravka. Positive results have been achieved in improving the properties of crosses. However, initiated work on the improvement of national pig breeding was short-lived (**Figure 2**).

2.3 Resavka

Resavka was created in the same area of Serbia but in much lower numbers than Moravka. It was reared especially in the valley of the river Resava. It was also formed by the non-systematic crossing of domestic breeds (Šumadinka and Mangalitsa) with Berkshire breed. On the one hand, single color black animals were chosen and, on the other, spotted animals, which resulted in stabilization of two types or pig breeds. The production characteristics of Resavka are similar to properties of the Moravka breed (**Figure 3**).

Figure 3.
Resavka (Foto Č. Radović).

3. The size of the population and body development of pigs of indigenous breeds

3.1 Mangalitsa

The size of the population has changed over the years (**Table 1**), indicating the absence of stability in the breeding of Mangalitsa. Based on the data from 2017, the calculated effective population size is N = 193.46, which means that the breed is highly endangered. The number of animals under productivity control is insignificant, which represents a limiting factor for the genetic improvement of this breed (**Tables 2** and **3**).

In critically endangered populations, there is a loss of genetic variability, and the level of inbreeding is increased. This results in an increase in the frequency of recessive genotypes (sometimes undesirable) which reduces vitality, resistance, fertility, and animal growth. The level of inbreeding is related to the actual and effective size of the population. With the increase in the effective size of the population, the level of inbreeding is reduced and vice versa.

The data on the official website (http://efabis.tzv.fal.de/) show the following numbers (Republic of Serbia) for female and male breeding animals of Mangalitsa, Moravka, and Resavka breeds in 2014: 247 and 35, 18 and 4, and 6 and 2, respectively. According to the rulebook on incentives for the conservation of animal genetic resources [6], incentives are given for breeding sows, breeding boars, and breeding gilts of autochthonous breeds of Mangalitsa, Moravka, and Resavka. One of the conditions for obtaining incentives was the selection control, certified by the local, regional, and national breeding organizations. On the other hand, the report and the results of the implementation of the breeding program in 2014 in Central Serbia (Institute for Animal Husbandry, Belgrade,

Year	2009	2012	2013	2014	2015	2016	2017
Population	1000	100	300	300	1000	2000	2000
Breeding females	600	90	203	247	698	1914	1480
Breeding males	50	2	24	35	21	50	50
Breeding females registered in the herdbook		90	153	247	290	388	523

The calculated effective size of the Moravka population for 2017 is No = 56.78, which means that the breed is highly endangered.

Table 1.
Mangalitsa population size (EFABIS, European farm animal biodiversity information system. Available from http://efabis.vet.agri.ee/).

Year	2009	2012	2013	2014	2015	2016	2017
Population	100	150	100	300	100	500	500
Breeding females	90	140	14	18	95	257	265
Breeding males	10	10	2	4	1	15	15
Breeding females registered in the herdbook			13	18		84	184

The calculated effective size of the Resavka population for 2017 is No = 41.79, which means that the breed is highly endangered.

Table 2.
Moravka population size (EFABIS, European farm animal biodiversity information system. Available from http://efabis.vet.agri.ee/).

Year	2009	2012	2013	2014	2015	2016	2017
Population	40	50	5	8	16	64	75
Breeding females	30	40	4	6	15	60	65
Breeding males	10	10	1	2	1	4	5
Breeding females registered in the herdbook						20	18

Table 3.
Resavka population size (EFABIS, European farm animal biodiversity information system. Available from http://efabis.vet.agri.ee/).

Body measure	White Mangalitsa		Swallow Belly Mangalitsa		Red Mangalitsa	
	Boar	Sow	Boar	Sow	Boar	Sow
Withers height (cm)	83	81	78	78	88	82
Chest circumference (cm)	155	155	140	150	154	150
Body length (cm)	96	97	95	98	104	98
Body weight (kg)	190	165	165	170	220	180

Table 4.
Body dimensions/measures of adult animals (2–3 years old) of the Mangalitsa breed [8].

2015 [7]) show that only 102 litters of all three indigenous breeds of pigs were under control. A small number of breeding animals are under control; there is no interest of breeders and local breeding organizations in fattening autochthonous pig breeds. The records are incomplete, animals without pedigree are purchased, and there are no standard and selection criteria for autochthonous breeds.

Body development of adult animals of Mangalitsa breed (**Table 4**) shows that sows and boars of Red Mangalitsa have a higher body weight at the age of 2–3 years than the other two strains. Animals of Swallow Belly strain are smaller in size than the White and Red Mangalitsa strain.

Table 5 shows the body dimensions of adult animals of the Moravka breed. The body weight of the boars and sows varies in a wide range (72–152 and 70–160 kg, respectively). The body is over 32 and 33% longer than the withers height.

According to research [10] the average weight of sows (age 3–5 years) was 77.67 kg and height of the withers 63.83 cm. The average fertility of sows was 7.20 piglets.

Body measure	Sows	Boars
Withers height (cm)	62.40	63.44
Chest circumference (cm)	107.02	112.20
Body length (cm)	82.55	84.64
Body weight (kg)	93.70	98.00

Table 5.
Body dimensions/measures of the sows and boars of the Moravka breed (older than 2 years [9]).

4. Reproduction performance

The average age of the primiparous Swallow Belly Mangalitsa gilts differs between the herds, the rearing conditions, nutrition, and care—from 615.6 days [11] up to 966.6 days [12]. In the herds of Swallow Belly Mangalitsa covered by the project, the age of all controlled sows at first partus (**Table 4**) ranged from 430 (1.18 years, Breeder A) to 588 days (1.61 years, Breeder B).

In 90% of cases, with White and Swallow Belly Mangalitsa, the duration of gestation period is 113–117 days [13]. The same author concludes that the duration of the gestation does not depend on the age of sows and the number of piglets in the litter. Also, Petrović et al. [12] find that the herd, sow age, and mating season have no significant influence on the duration of gestation.

There is a significant variation in the average duration of the reproductive cycle of Mangalitsa sows between the herds, age of the sow, and the parity (**Table 6**).

The fertility of sows expressed by the size of the litter at birth and weaning varies between the herds, the age of sows, and the seasons. The diet, based mainly on corn, limits the reproductive performance of sows [14]. The sows of White, Swallow Belly, and Red Mangalitsa breed in Hungary gave birth on average to 6.73 and reared 5.92 piglets [15].

Studies by domestic authors show that the fertility of Mangalitsa is between 1 and 12 piglets in the litter or an average of about 5 piglets. In 9 of the 14 years, the fertility of the Swallow Belly Mangalitsa was higher than 5.50 piglets per litter [2]. Swallow Belly Mangalitsa reared in extensive conditions gives birth in the first partus from 3 to 4 and in subsequent parities from 5 to 6 piglets [16]. Recent studies show that the average number of live-born piglets is below 5 (**Table 6**).

Trait	[11]	[12]		[17]	[7]
		Herd 1	Herd 2		
Duration of gestation (days)	115.37	115.34	114.36		
Duration of reproduction cycle (days)	212.10	209.78	182.14		
Number of live-born piglets	4.48	4.54	4.87	4.82	4.73
Total born piglets	4.76	4.93	5.16	5.32	4.96
Number of reared piglets	3.85	3.83	4.80	4.92*	4.48

**Number of piglets per weaned litter*

Table 6.
Variability of reproductive performance of swallow belly Mangalitsa.

5. Housing systems, nutrition, and reproduction performance

Farming of autochthonous breeds in the Republic of Serbia is mostly in an open system (in forests using the grazing system). Only in the period of farrowing and rearing of piglets are sows housed in the facility. The diet for lactating sows consists of corn, barley (2.5 kg/sows), and kitchen waste. During the year when acorn is available, pigs are not given the corn. Pigs consume plant mass in the forest, wild fruits, roots, insects, etc. Piglets are fed about 2 kg of food. Fattening is carried out from August to November by adding corn in the amount of 3–4 kg/animal. Barley is added to the diet if necessary. In the winter period, alfalfa is added to diet for sows. Other categories of pigs receive maize (ground or on Cobb), in addition to plant ingredients from the nature.

Pig nutrition in an extensive system of keeping involves feeding pigs in pastures, in the forests, and in the winter additionally with corn, barley, and other cereals. Pig grazing positively affects the development of the digestive tract of the animal, allowing them to consume larger amounts of food. This is of particular importance for the feeding of lactating sows. Grazing can meet the needs of pregnant sows, but not sows during lactation. Concentrated and voluminous nutrients (beetroot, potato, pumpkin, etc.) can be added to sows' diet during lactation.

Keeping pigs in the forest means that animals consume the natural food they find themselves. The pigs' nutritional requirements cannot be met only by what they find in the forest, but it can be combined with pastures and fields after harvest. This way of keeping is environmentally friendly as it provides the natural way of rearing indigenous breeds of pigs. Movement of pigs in forest and pasture areas provides a positive impact on the welfare, health of pigs, and the acquisition of quality raw materials for further processing.

Pigs that are reared in the forest mostly consume acorn (oak and beech), wild chestnuts, wild fruits, roots, insects, worms, etc. During the winter period, but also in the diet of pregnant and sows in lactation, grains and voluminous nutrients should be added. Also, with this rearing system, financial investments in the construction of expensive facilities and food costs are lower. The benefits of using acorn in pig nutrition are in its chemical composition and antioxidant properties. Acorn is rich in tocopherols and tannins. Feeding the pigs with acorns at the final stage of the fattening positively influences the fatty acid composition of the muscles. Beech acorn is nutritionally similar to corn and is considered good feed stuff in pig fattening. In a closed housing/rearing system of fattening, it can be crushed or ground. Oak acorn is less nutritious than beech because it contains more cellulose and less protein. Daily quantities of acorn per animal depend on the composition of the meal, i.e., the share of other nutrients. In the literature, daily amounts are reported from 3.1 to 3.6 kg [18] up to 4 kg [19] and from 7 to 10 kg of acorn [20]. Pigs also consume wild chestnut and sweet chestnut. The nutritional value of chestnut is similar to the acorn.

Reproductive properties of 192 Swallow Belly Mangalitsa sows and size of 536 litters in four breeders (A, B, V, and G) are shown in **Table 7**. The number of litter per sow ranged from 2.00 (Breeder A) to 3.70 (Breeder V). Breeder A in the parity structure has 57.81% of the firstborn, which means he is expending his herd. Contrary to Breeders A, B, and G, Breeder V has only sows with more than one parity (sows with 2–5 litters).

If the population is reproductive inactive and if the population includes fewer individuals per year, the population is compromised.

The average number of live-born piglets per litter ranges from 3.18 (Breeder B) to 4.46 (Breeder V). The fertility of all sows in four herds was 3.92 live-born piglets. The sows farrowed 1 to 7 piglets per litter. The difference in the average number of live-born piglets LBP (**Table 7**) between the breeders ranges from 0.38 (difference between A and V) to 1.28 piglets (difference between B and V). The share of stillborn piglets is in the range of 1.45 (Breeder A) to 9.14% (Breeder B). In the free rearing/housing system, sows are often separated from the herd, looking for a protected area to give birth to piglets, which are exposed to dangers and death. If farrowing is done within the facilities, it is possible to determine the share of stillborn piglets more reliably. The sows reared on average 3.76 pigs with a variation of 3.00 (Breeder B) to 4.29 (Breeder V). The difference in the average number of reared piglets between Breeder A and V (0.25 piglets) is not statistically significant. Breeders B and G have sown whose fertility is below the average in all four herds (**Table 8**).

The average fertility values of the primiparous females are shown in **Table 9**.

In the first partus, sows had an average of 3.54 piglets, of which 3.46 were live-born. The average number of reared piglets was 3.35 or 96.8% of the total born

Traits	**Breeder**[1]			
	A	**B**	**V**	**G**
Number of sows under control	64	38	20	70
Number of litters	127	131	74	204
Litters per sow	2.00	3.45	3.70	2.91
Share of first litters (%)	57.81	21.05	0	32.86
Age at first farrowing (A1F, years)	1.18	1.61	1.20	1.29
Number of live-born piglets (LBP)	4.08	3.18	4.46	3.62
Total born piglets (TBP)	4.14	3.50	4.65	3.76
Share of stillborn piglets (%)	1.45	9.14	4.09	3.72
Number of reared piglets (RP)	4.04	3.00	4.29	3.41
Piglet losses during lactation (%)	0.98	5.66	3.81	5.80

[1]Belgrade Region, Breeders A and G; Mačva Region, Breeder V; Moravica Region, Breeder B

Table 7.
Average values of fertility traits of swallow belly Mangalitsa sows (samples from the project).

Breeder	**Number of litters**	**LBP**	**TBP**	**RP**
A-B	-1.45^{***}	$+0.90^{***}$	$+0.64^{**}$	$+1.04^{***}$
A-V	-1.70^{***}	-0.38^{*}	-0.51^{**}	-0.25^{ns}
A-G	-0.91^{**}	$+0.46^{**}$	$+0.38^{*}$	$+0.63^{***}$
B-V	-0.25^{ns}	-1.28^{***}	-1.15^{***}	-1.29^{***}
B-G	$+0.54^{ns}$	-0.44^{*}	-0.26^{ns}	-0.41^{*}
V-G	$+0.79^{*}$	$+0.84^{***}$	$+0.89^{***}$	$+0.88^{***}$

$^{ns}P \geq 0.05$
$^{*}P \leq 0.05$
$^{**}P \leq 0.01$
$^{***}P \leq 0.001$

Table 8.
Significance of differences between the arithmetic mean of the fertility traits of the Swallow Belly Mangalitsa sows between the breeders.

Traits	Breeder			Average
	A	B	G	
Number of sows under control	37	8	23	68
Age at first farrowing (A1F, years)	1.25	1.80	1.32	1.34
Number of live-born piglets (LBP)	3.95	2.88	2.87	3.46
Total born piglets (TBP)	3.97	2.88	3.09	3.54
Number of reared piglets (RP)	3.86	2.88	2.70	3.35

Table 9.
Average values of fertility traits of primiparous Swallow Belly Mangalitsa sows (samples from the project).

Breeder	A1F (years)	LBP	TBP	RP
A-B	-0.55^{***}	$+1.07^{***}$	$+1.09^{**}$	$+0.98^{**}$
A-G	-0.07^{*}	$+1.08^{***}$	$+0.88^{***}$	$+1.16^{***}$
B-G	$+0.48^{**}$	$+0.01^{ns}$	-0.21^{ns}	$+0.18^{ns}$

$^{ns}P \geq 0.05$
$^{*}P \leq 0.05$
$^{**}P \leq 0.01$
$^{***}P \leq 0.001$

Table 10.
Significance of differences in arithmetical means of fertility traits of primiparous Swallow Belly Mangalitsa sows between the breeders.

piglets. The number of live-born piglets ranged between the litters on average from 2.87 (Breeder G) to 3.95 (Breeder A), total born piglets from 2.88 (Breeder B) to 3.97 (Breeder A), and reared from 2.70 (Breeder G) to 3.86 (Breeder A).

Fertility of primiparous sows of Breeder A was higher by 1.07 and 1.08 live-born piglets and 0.98 and 1.16 reared piglets than in the case of Breeders B and G (**Table 10**). The average size of the litter (at birth and weaning) of the primiparous sows of Breeders B and G was not significant.

6. Quality of carcass sides and meat of autochthonous breeds

6.1 The impact of the rearing system and nutrition

Mangalitsa is a typical fatty pig breed, which has about 30–35% of meat in carcass sides [4, 21–23]. Moravka is a breed of combined production abilities that has more meat in carcass sides and less fat than Mangalitsa [22, 23].Today autochthonous breeds are reared in an open system or farm conditions and fed in a traditional way or with complete feed mixtures.

Study of the influence of different methods of rearing and nutrition on the quality of carcass side and meat of Swallow Belly Mangalitsa was performed on 23 fatteners (male castrated animals, [24]). The first group was kept in an open system in the forest, mainly grazing and consuming roots and forest fruits with the addition of smaller quantities of corn (up to 0.3 kg/animal/day). The second group was reared in farm conditions with free ranges and fed with two complete feed mixtures (the first mixture with 15 and the other with 13% protein content). Animals were slaughtered in the same slaughterhouse. On the slaughter line, linear measures of warm carcass sides were taken, and the pH of the long *musculus longissimus dorsi* (MLD) was measured in the first hour after slaughter. A partial dissection of left chilled carcass sides was carried out according to the EU reference method [25].

The total mass and mass of muscle tissue in the four main carcass parts depended on the mass of cold carcass sides. The system of rearing and nutrition of Swallow Belly Mangalitsa fatteners showed no impact on most of the carcass side traits. However, it influenced the age of slaughterers at slaughter. The fatteners kept in the open system had an average of 739.0 days at slaughter, compared to 348.8 days in the closed system.

Of the total weight of the ham, the muscle tissue was 45.58 (open system) and 48.32% (closed system). A similar proportion of muscle tissue was found in the shoulder (47.90 and 45.67%). The share of muscle tissue in the back loin and belly rib carcass parts of fatteners reared in the closed system was higher (28.87 and 28.06%) than the open-system fattening (25.59 and 25.12%), but the differences were not significant ($R > 0.05$).

Fatteners kept in the closed system had by 2.53% more muscle tissue in carcass sides (mean 37.07%) than those reared in the open system (mean 34.54%); however, the established difference was not significant. MLD of fatteners reared in the closed system showed significantly higher water content (by 2.22%) and lower total fat (by 2.64%) than the open holding system. Statistically significantly higher ash content (by 0.07) was established in MLD of fatteners kept in the closed rearing system. The average pH_1 values indicate that both groups of fatteners had normal-quality meat with a higher variability of the trait in animals reared in the closed system, so that the difference of 0.28 was not significant.

6.2 The impact of breed, pig gender, and breeding methods

The study of the influence of the breed and gender of the fatteners on the growth traits, the composition of the carcass sides, and the quality of the meat was carried out on the animals reared in the same conditions [26]. The castrated males and females of the Swallow Belly Mangalitsa (LM, n = 19) and Moravka (M, n = 23) breeds were grown in the same herd in farm conditions. Each animal was provided with a surface area of 3.57 m^2. The fattening began with about 20–22 kg and lasted until animals reached 93–124 kg body weight. During the fattening, the animals were fed with two complete mixtures consisting of maize silage, livestock flour, soybean meal, sunflower meal, synthetic lysine, mineral nutrients, and premixes.

The average slaughter weight of LM and M was 103.67 ± 1.30 and 103.53 ± 1.21 kg and age 339.08 and 331.35 days, respectively. In the performed researches, the average daily gain from birth to the end of the fattening was lower and did not differ between LM and M (307 and 316 g, $P > 0.05$).

Fatteners of Moravka breed had a higher average weight of the back loin part of the carcass side (by 0.731 kg) and the weight of the subcutaneous fat tissue with skin (for 0.355 kg) than the Swallow Belly Mangalitsa breed (**Table 11**). The share of muscle tissue in this carcass part in the total weight of muscle tissue in four parts of the carcass side was higher in the animals of the Moravka breed (by 2.18%). The greater share of bone tissue in this part of the carcass side of the M breed led to the fact that 38.90% of total bone weight in four parts of the carcass side was in the back loin part. Contrary to this, of the total bone weight in four parts of the carcass side, 21.61% was shoulder bone tissue, which is 1.94% more than in Moravka. Moravka had more skin and subcutaneous fat tissue on average by 0.968 kg.

Male castrated animals had lower average weight of ham (by 0.665 kg, $P < 0.05$), lower weight of muscle tissue in the ham (by 0.387 kg, $P < 0.01$), and higher share of bone tissue (by 0.54% $P < 0.05$) than female animals.

Carcass side part	Trait	Swallow Belly Mangalitsa	Moravka	Difference
Back loin	Weight, kg	6.545	7.276	0.731**
	Skin and subcutaneous fatty tissue, kg	3.345	3.700	0.355*
	Bone tissue, kg	0.733	0.837	0.104*
	Muscle tissue—share in four carcass side parts, %	21.35	23.53	2.18**
	Bone tissue—share in four carcass side parts, %	35.81	38.90	3.09*
Shoulder	Bone tissue—share in four carcass side parts, %	21.61	19.67	1.94**
Four main carcass side parts	Skin and subcutaneous fatty tissue, kg	10.817	11.785	0.968**

*$P < 0.05$
**$P < 0.01$

Table 11.
Influence of breed on weight variation and tissue share in carcass side parts (only significant differences).

The share of muscle tissue in the ham in the total weight of this tissue in the four main parts of the carcass side was lower (by 2.31%) than in female animals. In the belly rib part of the carcass side, the castrates had less intermuscular fat tissue (by 0.187 g) and bone (by 32 g) than females. The intermuscular fatty tissue in the belly rib carcass part of the castrates made 29.44% of this tissue in four main carcass parts, which is 5.54% less than in females. The bone weight in the belly rib carcass part of the castrates was 11.33% and in females 12.70% of this tissue in four parts of the carcass side. A smaller share of the bone tissue in the castrates of 1.37% is statistically significant. The skin and subcutaneous fatty tissue in the back loin part made 32.82% of these tissues in four parts of the carcass side, which is by 2.81% more than in female animals.

Quantitative indicators of the quality of carcass and meat [27] were examined in fatteners of three genotypes and two Mangalitsa strains (White BM and Swallow Belly strain—LM of Mangalitsa breed) and Swedish Landrace (SL). In total, 36 male castrated animals were examined. Experimental animals were reared in the same facility from 20–25 to 100 kg body weight. Each animal was provided with 5 m^2 of surface. The diet was ad libitum with two complete mixtures. Animals were slaughtered at a body weight of about 100 kg, in the same slaughterhouse. The average slaughter weights of BM, LM, and SL were 100.7, 100.8, and 96.2 kg, respectively. There were no significant differences in slaughter body weight between genotypes. *Musculus longissimus lumborum et thoracis* (MLLT) SL contained more water (72.7%) than BM (64.3%) and LM (62.7%). The share of proteins in MLLT was significantly ($P < 0.001$) higher in SL (22.1%) and SBM (21.1%) fatteners than in LM fatteners (19.5%). Contrary to this, SL fatteners had less fat (4.23%) than BM and LM (13.5 and 16.8%). The established differences were significant at 99%. The content of saturated fatty acids (SFA) in 100 g of MLLT was higher in SL (43.4 mg) than in LM (35.3 mg) and BM animals (33.8 mg). There were no significant differences in the content of SFA between the strains of Mangalitsa. Contrary to this, the content of monounsaturated fatty acids (MUFA) was lower in the MLLT of castrates of the Swedish Landrace (44.9 mg) than LM (55.1 mg) and BM (58.0 mg) strains. The ratios of n-6/n-3 polyunsaturated fatty acid (PUFA) in BM, LM, and SL were 45.63, 14.05, and 34.01%, respectively. The share of cholesterol in MLLT was the lowest in SL fatteners (47.1% versus 62.3 and 62.9%).

Two autochthonous breeds (Swallow Belly Mangalitsa and Moravka) and crosses of these breeds (F1 generation) were reared under the same conditions and fed with mixtures of the same composition [28]. At an average age of 338 days, they had 110.40 kg. In the left carcass side, they had an average of 31.06 ± 3.73% of meat. Meat was of normal quality (pH_1 = 6.35 ± 0.24). The muscle tissue of the ham, shoulders, belly rib, and back loin part was 45.02, 43.26, 29.20, and 25.71%, respectively. Fatteners increased body weight by 329 g/day. The average daily gain in the weight of warm carcass sides was 271 g. The skin and subcutaneous fatty tissue in the four main parts of the carcass side on average showed more intense increase (37 g/day) than the muscle tissue (29 g/day). F1 generation crosses had more intensive body weight gain (352 g), weight of warm carcass sides (295 g), muscle tissue (32 g), and skin and subcutaneous fat tissue weight in four parts of the carcass side (41 g) than the average obtained for the parent breeds. The heterosis effect for calculated daily gain was 8.81, 11.3, 14.3, and 15.5%, respectively. A more intense increase in the weight of skin and subcutaneous fatty tissue is not preferable in four carcass parts of the F1 generation.

7. The fatty acid composition of the long back muscle

7.1 The influence of the housing system, breed, and pig gender

Oils and fats are essential in human nutrition; however, the health condition is negatively affected by too much or too little fat. The World Health Organization [29] reports a link between nutrition and chronic illnesses. In the human nutrition, there should be 15–30% of energy from fat, of which less than 10% should be saturated fatty acids because their higher levels increase the content of cholesterol and triglycerides in the blood. The share of polyunsaturated fatty acids should be 6–10% due to the need for essential fatty acids. Preferably, the *n*-6 and *n*-3 PUFA should be at 5–8% and 1–2%, respectively, but not more than 1% of trans-fatty acids. Most fatty acids are monounsaturated fatty acids (MUFA), primarily oleic acid (C18:1n-9). They are less susceptible to oxidation and have a positive effect on cholesterol levels. Increased intake of *n*-3 fatty acids in relation to *n*-6 has a positive effect on human health [30]. Pig meat is richer in linoleic acid (C18:2, *n*-6), which increases the total content of *n*-6 fatty acids in pork products [31]. The fatty acid composition of the pig fat and muscle tissue is influenced by a number of factors such as genetic factors [32–34], breed [35–38], gender, body weight [39], age, energy intake, fatty acid composition of the diet, and housing system [11, 36, 40, 41].

The results of the experiment [24] showed that the male castrated animals of Mangalitsa breed reared in the free system had more SFA (35.5 versus 33.9%) and *n*-3 PUFA (0.641–0.152%) than the fatteners in the conventional rearing system. The *n*-6/*n*-3 ratio in MLD was higher in fatteners reared in the conventional system than in the free system (37.3–9.2). The cholesterol content was not significantly different in MLD fatteners kept in different systems (61.7–63.1 mg/100 g). Fatteners reared in the free system had more proteins (21.7–19.0%) and less fat (12.1–18.2%) in the MLD than the animals in the conventional system.

The chemical composition [40] of the long back muscle (MLD) was examined in fatteners of Swallow Belly Mangalitsa and Moravka.

Body weight at slaughter of ML and M fatteners was on average 107.14 ± 2.85 and 107.61 ± 3.06 kg, respectively. Also, the slaughter weight of fatteners did not vary between castrates and gilts. A smaller share of water in MLD of Moravka animals meant that there was more total fat (6.96% compared to 5.10%) and the established difference was significant. The male castrated animals of M breed had more fat than gilts of the same breed (8.64–5.29%), and the difference in mean values of 3.35% was statistically significant. Significant differences between the mean values of fat content (5.25–4.94%) were not found between fatteners of different genders of the ML breed. The protein content of MLD of Swallow Belly Mangalitsa was increased by 0.92% compared to Moravka, and this difference was statistically highly significant.

In the study by Migdal et al. [42], no differences between Mangalitsa and Moravka in protein content (20.7–20.2%) were found. Meat proteins, which can be up to 24%, have a high biological value. The water and protein content are in a relatively constant ratio, i.e., there is 3.2–3.7 times more water in meat than protein, as stated by *Vuković* [43]. In our studies, this ratio is 3.2 (breed ML) and 3.3 (breed M). For fatty acid profile, Swallow Belly Mangalitsa was characterized by statistically the highest level of n-6 and n-3 PUFA (7.771 ± 0.728 and 0.416 ± 0.038, respectively), and CLA levels were the highest in Moravka and the lowest in Swallow Belly Mangalitsa (0.079 ± 0.010 and 0.072 ± 0,007, respectively) in the results of the experiment by Migdal et al. [44].

The breed of fatteners influenced the variance of total saturated fatty acids (ΣSFA) in MLD (**Table 12**) [40]. M breeders had an average of 41.64% and fatteners ML 39.45% of SFA. The difference in mean SFA of 2.19% is statistically significant. Both genders of ML fatteners had less saturated fatty acids than M fatteners.

The content of all monounsaturated fatty acids (ΣMUFA) varied under the influence of the breed. ML fatteners had more total monounsaturated fatty acids than M fatteners (56.41–53.78%). The difference in the corrected mean values of MUFA between breeds (2.63%) was statistically significant. Significant variations between male castrated and female animals of the same breed were not found. Male castrates and female animals of Swallow Belly Mangalitsa breed had higher content of MUFA (56.92 and 55.9%) than both genders of the Moravka breed (53.56 and 53.91%). The share of total polyunsaturated fatty acids (ΣPUFA) did not vary under the influence of the breed.

Trait	LS Mean ± SE	
	Mangalitsa	Moravka
ΣSFA	39.45 ± 0.55^a	41.64 ± 0.57^a
ΣMUFA	56.41 ± 0.56^a	53.78 ± 0.58^b
ΣPUFA	4.10 ± 0.30	4.54 ± 0.31
C14:0	1.33 ± 0.03	1.34 ± 0.03
C16:0	25.05 ± 0.32	25.53 ± 0.34
C16:1	4.19 ± 0.14^a	3.70 ± 0.15^a
C17:1	0.34 ± 0.02^a	0.24 ± 0.03^a
C18:0	12.73 ± 0.37^a	14.40 ± 0.39^b
C18:1	50.82 ± 0.48^a	48.51 ± 0.50^b
C18:2	3.92 ± 0.28	4.26 ± 0.29
C18:3	0.21 ± 0.04	0.31 ± 0.04
C20:0	0.23 ± 0.04	0.26 ± 0.05
C20:1	1.07 ± 0.08^a	1.32 ± 0.08^a
P/S†	0.10 ± 0.01	0.11 ± 0.01

$^{a-b}P < 0.05$

Table 12.
Influence of fatteners' breed on variation of fat acid composition/profile of musculus longissimus dorsi.

The most common saturated fatty acids in MLD of Mangalitsa and Moravka fatteners were C16:0 (palmitic acid, 25.05 and 25.53%) and C18:0 (stearic, 12.73 and 14.40%). Both fatty acids C14:0 and C20:0 made 4.0 and 3.8% of all SFAs. The breed influenced the variation of the stearic acid content. In MLD of fatteners M, there was by 1.67% more stearic acid than ML.

The most common monounsaturated fatty acid was C18:1 (oleic). Fatteners of Swallow Belly Mangalitsa had more oleic acid than Moravka (M, 50.82 versus 48.51%). The content of C18:1 in MLD did not vary between genders of the same breed. The second common MUFA was C16:1. Significant differences in palmitoleic acid mean values (0.49%) were found between LM and M fatteners. The content of eicosenoic acid in MLD varied between the breeds but did not vary statistically significantly between the genders of the same breed. MLD of M fatteners contained more C20:1 by 0.25%. The variation of the content of C17:1 was influenced by the breed of fatteners, so MLD in Mangalitsa fatteners had more of these fatty acids (by 0.10%) than in the Moravka breed. The most common polyunsaturated fatty acids were C18:2 (linoleic acid) compared to C18:3 (linolenic acid). The contents of C18:2 and C18:3 in MLD did not vary under the influence of breed and gender of fatteners.

If the obtained mean values for fatty acids in MLD of Mangalitsa, Moravka, and meaty breed (SL), the results from our previous research [35] reared under the same conditions, are compared, it can be concluded that ML and M had a lower content of SFA (39.45, 41.64, and 43.76%, respectively), higher content of MUFA (56.41, 53.78, and 41.22%, respectively), and lower content of PUFA (4.10, 4.54, and 14.74%, respectively). Fatteners of SL breed (Swedish Landrace) compared to the ML and M breeds had more total SFAs (4.31 and 2.12%), fewer MUFAs (15.19 and 12.56%), and more PUFAs (10.64 and 10.20%).

Polyunsaturated essential fatty acid (linoleic acid, C18:2 n-6) introduced to the organism through food passes through the digestive tract of the pig unchanged; it is resorbed from the small intestine to the bloodstream and is incorporated into the tissue. Pig meat is richer in linoleic acid content (C18:2, n-6), which increases the total content of n-6 fatty acids in pork products [31]. The pig nutrition in the final stage of fattening which includes acorn positively influences the fatty acid composition of the muscle, i.e., the content of oleic acid was higher and the shares of palmitic and stearic acid lower. Also, nutrition ad libitum in the last 3 weeks has affected the share of linoleic acid in MLD, so the n-6/n-3 fatty acid ratio was three times lower in black Slavonian pigs than in pigs that were fed with complete mixture. There were fewer triglycerides (48%) and cholesterol (by 11%, [45]) in the blood of pigs fed with acorn (ad libitum in the last 3 weeks). Meat of pigs reared in this way is better for processing into traditional meat products but also has better nutritional value for human consumption.

Grazing/pasture is the source of n-3 fatty acids. Pig nutrition on pasture leads to the increase in the levels of linoleic and omega-3 fatty acids and reduction in the ratio of omega-6 to omega-3 fatty acids [46].

The ratio between polyunsaturated and saturated fatty acids should be greater than 0.4, but on the other hand, not only the high content of polyunsaturated fatty acids is sufficient, but also the n-6/n-3 ratio is important. It is recommended that the ratio between PUFA/SFA should be greater than 0.45 and lower than 1.0. In our research, the P/S ratio was not favorable because it was 0.10 (ML) and 0.11 (M).

Meat of two indigenous breeds (Swallow Belly Mangalitsa and Moravka) and meaty breeds was used for the production of Kulen sausage [23, 47]. In all types of Kulen sausage, there was on average 33.75% protein. The protein content in Kulen sausages A and J (made using the meat of Mangalitsa pigs) was 34.62 and 27.18%, respectively. More than 35% protein was in Kulen B (made from meat of meaty breeds, 35.79%), D (made from meat of all three pig breeds, 35.63%), and E (made using the meat of Moravka pigs, 35.04%). The contents of proteins, cholesterol, saturated fatty acids, monounsaturated fatty acids, and polyunsaturated fatty acids in Kulen sausage S were 34.42, 66.00, 40.21, 45.79, and 14.00%, respectively. Kulen sausage S (produced from 70% of Mangalitsa meat and 30% Moravka meat) had the highest average mean score for all eight organoleptic properties (5.20 ± 0.49) given by professional assessors. Kulen sausage S had an average ranking of 1.38 ± 0.96 in terms of overall acceptance, which means that professional appraisers and appraisers-consumers had the same choice.

8. Conclusion

In our country, as well as in surrounding countries, autochthonous pig breeds are reared in an open and closed system (farm conditions) and fattened to a different final body weight depending on market demands. On the one hand, the number of animals of Mangalitsa and especially Moravka breed is low, and the production is unorganized. On the other hand, meat products of autochthonous pig breeds have a high price and are available to a small number of consumers.

Potentially endangered indigenous breeds of pigs in Serbia are registered, phenotypically defined, and recorded in appropriate databases. Previous results have shown that there are problems relating to their identification, records, control of production traits, planned mating, coefficient of relation between them, and inbreeding within autochthonous races. For a safer and more justified conservation program, a more accurate, more reliable characterization of pig breeds on the list of protected (Mangalitsa, Moravka, and Resavka) is necessary. Using the method of molecular genetics, the status of authenticity or autochthonousness of pig breeds, which are included in the program of conservation, must be established and carried out with a number of farmers—animal owners.

Preliminary results of the genetic characterization of breeds with MS markers, using two programs (Faculty of Agriculture, University of Osijek), clearly separate animals of Mangalitsa breed from Moravka. They also show that Moravka breed is less uniform and consists of at least two populations. The reason may be the possible crossings that have occurred. Literary data show that there were two types of Moravka. About 91% of the blood samples analyzed (Breeder A) show that these animals belong to Mangalitsa breed. In the herd of the Breeder G, only 54% of the analyzed samples show affiliation with Mangalitsa breed, and 46% have genes of other breed/breeds.

After determining the genetic affiliation of the animal to the endangered breeds, herds are to be formed, which should be permanently protected in order to preserve the specificity of the genetic value. Breeds that are not intended for breeding should be included in commercial programs or organic/ecological production in order to obtain raw materials for the production of traditional products that would have significance in rural development.

In the coming period, the focus in in situ conservation, population increase, and research should be on Resavka breed.

Acknowledgements

This review research was financed by the Ministry of Education, Science and Technological Development of the Republic of Serbia, project TR 31081.

Author details

Čedomir Radović[1*], Milica Petrović[2], Marija Gogić[1], Dragan Radojković[2], Vladimir Živković[1], Nenad Stoiljković[1] and Radomir Savić[2]

1 Institute for Animal Husbandry, Belgrade, Republic of Serbia

2 University of Belgrade Faculty of Agriculture, Belgrade, Republic of Serbia

*Address all correspondence to: cedomirradovic.izs@gmail.com

References

[1] Bogdanović SV. Mangulica, rasa srpskih svinja. Težak. 1886;**XVII(8)**:475-477

[2] Belić J, Gajić Ž, Isakov D, Ognjanović A, Šterk V. Savremeno Svinjarstvo. Beograd: Privredni pregled; 1972. pp. 1-621

[3] Nikolić D. Svinjarstvo. Udžbenik za srednje pšoljoprivredne škole; 1951. pp. 1-252

[4] Egerszegi I, Ratky J, Solti L, Brussow KP. Mangalica—An indigenous swine breed from Hungary. Archiv fur Tierzucht. 2003;**46**:245-256

[5] Zsolnai A, Radnóczy L, Fésüs L, Anton I. Do Mangalica pigs of different colours really belong to different breeds? Archiv fur Tierzucht. 2006;**49**:477-483

[6] Republic of Serbia. Official Gazette of the Republic of Serbia. 2013. No. 83/2013

[7] Radović Č. Annual Report. Belgrade, Serbia: Institute for Animal Husbandry; 2015

[8] Szabó P. The Mangalica. (a mangalica). Kistermelők Lapja. 1999;**12**:14-15

[9] Živković R, Kostić J. Prilog poznavanju crne i šarene svinje (moravke i resavke). Arhiv za poljoprivredne nauke. 1952;**V**(10):23-46

[10] Petrović M, Mijatović M, Radović Č, Radojković D, Josipović S. Genetic resources in pig breeding—Carcass quality traits of breeds Moravka and Mangalitsa. Biotechnology in Animal Husbandry. 2007;**23**(5-6):421-428

[11] Parunović N, Petrović M, Matekalo-Sverak V, Trbović D, Mijatović M, Radović Č. Fatty acid profile and cholesterol content of *M. longissimus* of free-range and conventionally reared Mangalitsa pigs. South African Journal of Animal Science. 2012;**42**:101-113

[12] Petrović M, Savić R, Parunović N, Radojković D, Radović Č. Reproductive traits of pigs of Mangalitsa breed. In: 8th International Symposium on the Mediterranean Pig, Slovenia, Ljubljana; 10-12 October 2013; Acta argiculturae Slovenica, Supplement; 2013;**4**:89-92

[13] Živković S. Prilog poznavanju trajanja suprasnosti kod nekih rasa svinja. Stočarstvo. 1956;**X**(5-6):234-240

[14] Tóth S. Heridity of prolificacy, litter-weight on the 30 and 60 days in two Mangalica stocks. (A szarvasi és a mezőhegyesi mangalica állomány szaporaságának, 30 és 60 napos alomsúlyának örökölhetősége). Állattenyésztés. 1962;**11**(1):43-46

[15] Szabó P. Theriogenological results of alternative pig breeds. (Alternatív sertésfajták szaporodásbiológiai eredményei). In: Innováció, a tudomány és a gyakorlat egysége az ezredforduló agráriumában. Debrecen; 2002. pp. 97-102

[16] Brinzej M. O Lasastoj mangulici s obzirom na njen uzgoj u šumi. Stočarstvo. 1948;**12**:293-298

[17] Petrović M. Annual Report. Belgrade, Serbia: University of Belgrade; 2010

[18] Rodriguez-Estavez V, Sanchez-Rodriguez M, Garcia M, Gustavo Gomer-Castro A. Feed conversion rate and estimated energy balance of free grazing Iberian pigs. Livestock Science. 2010;**132**:152-156

[19] Budimir K, Margeta V, Steiner Z, Crnac V. Hranidba svinja u ekstenzivnim uvjetima držanja. In: 50 Croatian and 10 International Symposium on

Agriculture. Opatija, Croatia; 2015. pp. 414-417

[20] Canellas I, Roig S, Poblaciones MJ, Gea-Izquierdo G, Olea L. An approach to acorn production in Iberian dehesas. Agroforesty system. 2007;**70**:3-9

[21] Kralik G, Petričević A. Production traits of black Slavonian pig. In: Proceedings of the Biological Diversity in Animal Production of Republic of Croatia, Zagreb; 2001. pp. 115-122

[22] Petrović M, Mijatović M, Radović Č, Radojković D, Parunović N, Stanišić N. Genetic resources in pig breeding—Carcass and meat quality traits of moravka and mangalitsa breeds. In: 1st Conference of the Balkan Network for the Animal Reproduction Biotechnology Institute of Biology and Immunology of Reproduction. Acad. Kiril Bratanov. 2009;14

[23] Petrović M, Radović Č, Parunović N, Mijatović M, Radojković D, Aleksić S, et al. Osobine kvaliteta polutki i mesa svinja rase moravka i mangulica. Biotechnology in Animal Husbandry. 2010;**26**(1-2):21-27

[24] Petrović M, Radović Č, Parunović N, Radojković D, Savić R. Composition of carcass sides and quality of meat from swallow-belly Mangalitsa reared in two systems. Biotechnology in Animal Husbandry. 2012;**28**:303-311

[25] Walstra P, Merkus GSM. Procedure for assessment of the lean meat percentage as a consequence of the new EU reference dissection method in pig carcass classification. Report ID-DLO 96.014, March, 1996;22

[26] Petrović M, Radović Č, Mijatović M, Radojković D, Stanišić N, Parunović N. The shane of tissues in pig carcass sides of autochthonous breeds depending on the body mass and sex. In: 3rd international congress "new perspectives and challenges of sustainable livestock production"; 5-7 October 2011; Belgrade; 2011. Biotechnology in Animal Husbandry. **27**(3):561-569

[27] Parunović N, Petrović M, Matekalo-Sverak V, Radović Č, Stanišić N. Carcass properties, chemical content and fatty acid composition of the *musculus longissimus* of different pig genotypes. South African Journal of Animal Science. 2013;**43**(2):123-136

[28] Radović Č, Petrović M, Parunović N, Radojković D, Savić R, Stanišić N, et al. Carcass and pork quality traits of indigenous pure breeds (Mangalitsa, Moravka) and their crossbreeds. Indian Journal of Animal Research. 2017;**51**(2):371-376

[29] WHO/FAO. Diet, Nutrition and the Prevention of Chronic Diseases. WHO Technical Report Series 916, Report of a Joint WHO/FAO Consultation, World Health Organization and Food and Agriculture Organization of the United Nations, Geneva; 2003. Available from: http://www.fao.org/docrep/005/AC911E/ac911e00. htm#Contents [Accessed: 11 June 2014]

[30] Simopoulos AP. The importance of the ratio of omega-6/omega-3 essential fatty acids. Biomedicine and Pharmacotherapy. 2002;**56**:365-379

[31] Wood JD, Enser M, Fisher AV, Nute GR, Sheard PR, Richardson RI, et al. Fat deposition, fatty acid composition and meat quality: A review. Meat Science. 2008;**78**:343-358

[32] Sellier P. Genetics of meat and carcass traits. In: Rothschild MF, Ruvinsky A, editors. The Genetics of the Pigs. New York, NY, USA: CAB International; 1998. pp. 463-510

[33] Sellier P, Maignel L, Bidanel JP. Genetic parameters for tissue and fatty acid composition of backfat, perirenal

fat and longissimus muscle in large white and landrace pigs. Animal. 2010;**4**:497-504

[34] Fernández A, De Pedro E, Núnez N, Silió L, García-Casco J, Rodríguez C. Genetic parameters for meat and fat quality and carcass composition traits in Iberian pigs. Meat Science. 2003;**64**:405-410

[35] Parunović N, Petrović M, Matekalo-Sverak V, Radojković D, Vranić D, Radović Č. Cholesterol and total fatty acid content in m. longissimus dorsi of Mangalitsa and Swedish landrace. Acta Alimentaria. 2012;**41**:161-171

[36] Luković Z, Kaić A, Škorput D, Karolyi D. Effect of breed and rearing system on intramuscular fatty acid profile of M. semimembranosus in raw Slavonian ham. Italian Journal of Animal Science. 2009;**8**:255-257

[37] Daza A, Menoyo D, Olivares A, Cordero G, López-Bote CJ. Effect of Iberian pig feeding system on tissue fatty-acid composition and backfat rheological properties. Journal of Animal and Feed Sciences. 2007;**16**:408-419

[38] Furman M, Malovrh Š, Levart A, Kovač M. Fatty acid composition of meat and adipose tissue from Krškopolje pigs and commercial fatteners in Slovenia. Archiv fur Tierzucht. 2010;**53**:73-84

[39] Žemva M, Malovrh Š, Kovač M. Kakovost mišične in podkožne maščobe krškopoljskega prašiča in komercialnih pitancev. U: Kovač M., Malovrh Š, Urednici. Spremljanje Proizvodnosti prašičev, V. del. Domžale: Univerza v Ljubljani, Biotehniška fakulteta, Enota za prašičerejo, biometrijo in selekcijo; 2009. pp. 109-124

[40] Rey AI, López-Bote CJ, Buckley DJ. Effect of feed on cholesterol concentration and oxidation products development of longissimus dorsi muscle from Iberian pigs. Irish Journal of Agricultural and Food Research. 2004;**43**:69-83

[41] Petrović M, Wähner W, Radović Č, Radojković D, Parunović N, Savić R, et al. Fatty acid profile of *M. longissimus* dorsi of Mangalitsa and Moravka pig breeds. Archives Animal Breeding (Archiv Tierzucht). 2014;**57**(17):1-12

[42] Migdał W, Radović Č, Živković V, Gwiazda E, Migdał Ł, Migdał A, et al. Quality of meat from native pigs. In: Proceedings of the 11th International Symposium Modern Trends in Livestock Production; 11-13 October 2017; Belgrade 2017. pp. 189-203

[43] Vuković KI. Osnove Tehnologije Mesa. Beograd: Veterinarski fakultet; 1998

[44] Migdał Ł, Radović Č, Živković V, Gogić M, Migdał A, Migdał W. Characterization of meat traits and fatty acids profile from swallow-belly Mangalitsa, Moravka pigs and their crossbreeds. Annals of Warsaw University of Life Sciences—SGGW, Animal Science. 2018;**57**(4):365-378

[45] Salajpal K, Karoly D, Đikić M, Kantura V, Kiš G, Sinjeri Ž. Influence of acorn intake on blood lipid profile and longissimus muscle characteristics of black slavonian pig. Acta agriculturae Slovenica. 2008;**2**:99-105

[46] Nilzén V, Babol J, Dutta PC, Lundeheim N, Enfält AC, Lundström K. Free range rearing of pigs with access to pasture grazing—Effect on fatty acid composition and lipid oxidation products. Meat Science. 2001;**58**:267-275

[47] Parunović N, Petrović M, Matekalo-Sverak V, Radojković D, Radović Č. Fatty acid profiles, chemical content and sensory properties of traditional fermented dry kulen sausages. Journal of Food Processing and Preservation. 2014;**38**:2061-2068

Section 4

Treatments Used in Food Processing

Chapter 7

High Hydrostatic Pressure Treatment of Meat Products

Rosa María García-Gimeno and Guiomar Denisse Posada Izquierdo

Abstract

High hydrostatic pressure (HHP) treatment has been described to improve the microbiological safety and shelf life of ready-to-eat (RTE) meat products, as a nonthermal decontamination technology in the meat industry, applied at pre- or post-packaging. The pathogen widely studied in this product is *Listeria monocytogenes* that reflects the concern of the food industry. In general, microorganism's lethality during HHP treatment depends on specific intrinsic factors of the microorganism; those factors are related to food and technological factors of treatment. In addition to processing parameters, intrinsic factors of the food matrix also exert an effect on bacteria inactivation during pressure treatment. It is known that low water activity (a_w) protects microorganisms against the effects of pressure. Predictive modelling is an important tool of the novel microbial food safety management strategy that provides with accurate information to demonstrate and guarantee the safety and shelf life of the food products. The chapter describes the effect of parameters on the efficiency of this technology on meat products over pathogens, composition and the sensorial quality consequences. The predictive modelling tool is introduced for the optimisation of meat treatment.

Keywords: high hydrostatic pressure, high pressure processing, microbial inactivation, extension shelf life, predictive microbiology

1. Introduction

Today the food security situation is continually under review and questioned as a result of several food-borne outbreaks that occurred. The company, mainly responsible for the safety of its products, strives to achieve techniques and procedures that allow it to ensure all risks and at the same time extend commercial shelf life and all this without altering the sensory characteristics of optimum quality. Consumers demand insistently fresher, healthier, safer and more convenient food, with good tasting and without preservatives.

In the case of ready-to-eat products (RTE), the need is even more pressing since it is a product for direct human consumption without the need for cooking or other processing effective to eliminate or reduce to an acceptable level of microorganisms.

In this sense, European legislation (Regulation EC 2073/2005, [1]) establishes clear limits of various pathogens, such as regulating the presence of *L. monocytogenes* in ready-to-eat products, *Escherichia coli* in fruits, vegetables and live bivalve molluscs or *Salmonella* in ready-to-eat foods containing raw egg.

The occurrence of food-borne outbreaks in Europe has a decreasing tendency. A total of 5079 food-borne (including waterborne) outbreaks were reported by the European Food Safety Authority (EFSA) [2]. This report describes *Salmonella* as the commonest detected agent, and the highest-risk agent/food pairs the *Salmonella* in eggs and meat and meat products, and the analysis of strong-evidence food-borne outbreaks is associated with animal origin food [2]. In the case of meat products with longer shelf life, bacteria have more time to grow if they have the conditions (such as cooked sausages, cooked sliced ham and fermented salami) [3].

All the facts mentioned above has made companies look for alternative techniques that guarantee the safety of their products, as is the case of HHP. HHP has become a reality in the food industry and has spread worldwide [4]. This technique achieves a microbial inactivation without using high temperatures, so they manage to keep the sensory characteristics of the product almost intact, providing a larger commercial shelf life [5–7]. This preservation technique consists of the application of isostatic pressures, transmitted to foods uniformly and instantaneously by air-driven pumps through a liquid, generally water [8].

One of the main advantages of high pressure processing (HPP) is that it reaches acceptable microbial inactivation in meat products, but the sensorial and nutritional characteristics remain with good quality [9, 10].

In the meat industry, the application of HPP has focused on products ready for consumption with the additional aim of extending commercial life. For example, several studies have described the behaviour of *L. monocytogenes* in ready-to-eat meat products treated by HHP at different points of processes: prepackaging (liquid food) [11] and post-packaging (all types of food) [5, 12–19], the latter application being the most used [20]. In general, pathogen lethality during HHP treatment depends on various processing parameters such as the pressure level and holding time, temperature and food matrix. The optimisation of these parameters of the treatment for the pathogens' inactivation has been reinforced by the use of predictive microbiology tool that has been applied on different meat products [5, 15, 21–26].

Different organisations and administrations have recognised the listericidal effect of HHP treatments on RTE foods [27–29]. The objectives of this study are the revision of the effect of all parameters on the efficiency of this technology on meat products over pathogens and the sensorial quality consequences. Also, the predictive modelling tool for the optimisation of the meat treatment will be introduced.

2. High hydrostatic pressure treatment of meat

HHP was very well accepted since its beginning as an alternative to thermal inactivation treatments and as an in-package cold pasteurisation process [4]. In the last three decades, the number of companies with HHP facilities has increased considerably in the world, from just a few in 1990 to more than 200 units and with an increasing capacity [30].

The inactivation of bacteria effect of high pressure was demonstrated 100 years ago, although the industrial technology was built up at the end of the twentieth century [20]. The system consists in exerting high and uniform pressure on the food, for enough time to achieve the desired effect. This is called adiabatic heat and occurs instantaneously with pressure increase and as the pressure is uniform over the product [20]. HHP is probably the most developed nonthermal technology commercially in the world market, mainly applied for sliced meat products, fruit jellies and jams, fruit juices, dressings for salad, raw oysters, ham and guacamole, among others.

The packaged food is usually submerged in water inside a tank, and through this, high pressure is caused. HHP treatment will inactivate bacteria, yeast, moulds and enzymes equivalent to thermal pasteurisation processed but preserving the taste, colour and nutritious value of the product [4, 6, 23, 30, 31]. The treatment can be prepackaging (liquid food) or post-packaging (all types of food), although the latter is most frequently used [20]. The meat industry tries to apply the shortest HHP treatment on production lines as they can, currently from 3 to 6 min maximum [9, 32]; although many potential HPP applications would require long treatment times to ensure an adequate inactivation level of pathogens and spoilage microorganisms, pressure treatments alone would not be sufficient to guarantee food safety [33].

The application of HHP technology follows *two basic principles*: Le Chatelier principle and isostatic rule (Pascal principle). The first principle postulates that pressure accelerates reactions (phase change, changes in the molecular configuration, chemical reactions) that involve volume reductions and vice versa and inhibits reactions that occur with increases in volume. Since the medium used to transmit the pressure is usually water (incompressible fluid), the isostatic rule principle is verified in the HHP application, stating that "the increase in pressure applied to the surface of an incompressible fluid, contained in an undeformable container, is transmitted with the same value to each of its parts". The applied pressure is transmitted in an isostatic (uniform) and almost instantaneous way to all points of the food, regardless of its composition, size and shape. This prevents deformation of the product, despite being subjected to such high pressures, and makes it very homogeneous and does not have over-treated areas. When food is treated in its packaging, it must be flexible and deformable (it must tolerate volume reductions of up to 15%). The evacuation of gases from the interior is especially necessary to prevent their compression from reducing the pressurisation efficiency [34, 35].

The pressure range for a commercial purpose is usually from 100 to 600 Mpa [30], but this only can approach to a pasteurisation process but not commercial sterilisation where spores should be destroyed and more than 1000 Mpa should be applied for the sterilisation [30]. The new commercial unit implemented has increased in capacity and pressure, reducing the time to a few minutes, which helps manufacturers to reduce costs. The consumer has demonstrated a high level of acceptance of products treated with HHP because of the minimal changes in sensory and safety characteristics they perceive.

Two types of HHP treatment can be distinguished: the classical or also named single-pulsed HHP or the multi-pulsed high hydrostatic pressure (mpHHP). Difference between both is the number of compressions done. In the single HPP treatment, a compression hold for a certain time is followed by decompression to atmospheric pressure, while in the mpHHP more than one compression is applied with its respective decompression phase. It was reported that the mpHHP treatment, with few exceptions, is more effective than the classical or single-pulsed HHP treatment for inactivation of microorganisms in fruit juice, dairy products, liquid whole egg, meat products and seafood [4]. The reports of applying mpHHP on meat products describe better inactivation rates of *E. coli* O157:H7 and *Salmonella* Enteritidis in ground beef and chicken fillets, respectively, than the classic HHP [4, 36, 37]. Moreover, the mpHHP treatment could also be used to inactivate enzymes in foods and to increase the shelf life of foods [4].

The high pressure applied causes a temperature increase in the treated product, around 3°C per 100 MPa applied for water and 8–9°C for fat and oils and intermediated values for proteins and carbohydrates have been described [30, 38].

2.1 Effect of HHP on food components

The effectiveness of HHP on meat products constituents depends on different factors as initial microbial, pH and ionic strength [39].

The pressure affects properties of water contained in food such as density, viscosity, dipole moment, dielectric constant, and surface tension and thermal properties such as freezing and melting point and consequently will exert its effect on enzymes, chemical reactions and microorganism [40, 41]. For example, high pressures reduce the freezing point of water to −22°C at a pressure of 207.5 MPa because it prevents the increase in ice volume [40].

Whether the fat is affected or not by the treatment is very important because it has a significant impact on the sensory characteristics and it will depend on the intensity of treatment. It has been described that 450 MPa is applied during 154 s in dry fermented sausage; the total fatty acids and the stability of the fatty ones were not affected [42].

The denaturation that the proteins of the food undergo by the treatment of high pressures will depend on the level of pressure exerted, the pH and the temperature. Irreversible changes that have been described include the dissociation of oligomeric proteins into their subunit, aggregation or gelation of protein or changes in the conformation of the active site of enzymes. Reversible changes are observed when the pressures are between 100 and 300 MPa [30]. Proteins and sugars have been described as protective agents for bacteria in these treatments [5, 43–45].

2.2 Effect of HHP on the sensory quality of food

The effect of HHP on the sensory quality of food depends on the conditions, pressures and time, but physical properties of the food play an important role in its sensory quality. The colour of meat is critical because it is the main criterion that consumers will evaluate before making purchases.

The significant change of the texture and visual appearance, colour in the raw meat, depends on the intensities of pressure, observing significant changes at HPP at 600 MPa, but not at lower as 175 MPa [46, 47]. Nevertheless, on cases of cured meat products, changes on colour mainly depended on water content and water activity [48].

In case of salted chicken meat, it has been described that, in general, the use of HHP treatment improved the texture of cooked meat and colour of raw meat, and it is proposed as a processing alternative to reduce NaCl content [49]. Siddig et al. [50] in other study concluded that the colour of chicken was slightly affected by treatment, but pH, moisture content and the oxidation of lipids were not substantially changed.

Pressure treatment of meat can promote oxidation reactions, and it is crucial to control the balance between pro- and anti-oxidants to prevent this phenomenon because it will affect the colour. Lipid oxidation has been extensively investigated in meat because it can react with proteins, leading to organoleptic modifications and the loss of nutritional value. In the case of meat, the oxidation is one of the most important mechanisms of the degradation of meat, which can be initiated endogenously via metallic ions, especially hemic iron, or via exogenous reactive oxygen species. This process will result in changes in the organoleptic properties of the meat, as degradations in colour, aroma and flavour. These effects will be related to the type of meat, the treatment used and the methods used to evaluate the reactions with the oxidation of lipids and proteins. The pressure above 400 MPa seems to be critical for the initiation of lipid oxidation [7].

2.3 Effect of HHP on microorganisms

The effect of inactivation of HP on microorganisms in foods will depend on specific intrinsic factors of the microorganism, those related to food and technological factors of treatment [40, 51].

Among the intrinsic factors of microorganisms that will affect inactivation would be the number, species, strain and their physiological state [40, 51, 52]. Even the size of the microorganism has been described as influential [53]. In the different phases of physiological state, the cell and the membrane vary, and it has been observed that in the logarithmic phase of growth, it is more sensitive to the treatment of HHP and, in the stationary phases, it is more resistant [53].

The spores are even more resistant, and heat needs to be applied at the same time to inactivate them [34, 54]. For example, the spores of yeast and moulds had been reported to be inactivated by pressures of 600 MPa [9] although some species have been described as more resistant, as the ascospores of *Byssochlamys nivea* [40].

The factors related to food that affects the efficiency of treatment would influence variables such as pH, a_w, salt concentration and the general composition of the food [40, 51, 55].

The treatment gains effectiveness by lowering the pH of the food [52] or adding antimicrobials [34]. In a study by Alfaia et al. [42] carried out in chorizo, it describes a significant increase in pH by increasing the intensity of the treatment, which was also found in other products such as raw sausage batter, fresh chicken breast fillets and raw poultry sausages [26, 56]. At high pressure, there is increased ionisation and redistribution of ions that can be the origin of the pH increase and also the release of imidazolium groups by histidine [57]. Alfaia et al. [42] verified that the HPP resulted in a significant increase ($p < 0.001$) of the pH of chorizo compared to the control samples and in a significant decrease of the a_w ($p < 0.01$). The increase in pH was also reported on raw sausage batter, fresh chicken breast fillets and raw poultry sausages.

It has been observed that the decrease of a_w decreases the effectiveness of lethality of bacteria [4, 19], probably related to the stabilisation of protein, especially enzymes, which suffers less pressure [58]. It has been demonstrated that lyophilised *L. monocytogenes* treated with HPP was not inactivated [59]. On the other hand, it is also described that low a_w will inhibit the recovery of cells and potential growth during storage of the product treated by HPP [15, 60], that is to say that two antagonistic effects that could compensate each other.

Synergistic effects of HHP treatment with the addition of sodium lactate on the inactivation of *L. monocytogenes* in cooked chicken have been described [11].

Also, the fat content has been described as a parameter that affects the effectiveness of microorganism inactivation, having in general a protective effect of bacteria [5, 15, 19, 25]. High fat concentration decreases the inactivation of bacteria [15, 25], but it is also related to the pressure exerted; the higher the pressure of 650 MPa, the more is the protection [5, 18].

HPP and the addition of essential oils have similar effects on microbial structures, and thus they may act synergistically on the inactivation of microorganisms. Therefore, the combination of HPP with EOs is a promising alternative to expand the HPP food industry [61, 62].

The concentration of other components has been described affecting inactivation of bacteria as vitamins and amino acids [43], proteins [63], sucrose [64] and minerals such as calcium or magnesium [65].

Not only food component can affect the efficacy of HPP but also the food structure. Several authors have described it as an essential factor of variability on the resistance of microorganisms by comparing inactivation on food matrix and culture media where the food displays a protective effect against HHP [10, 66, 67].

Among the technological or process factors, the pressure exerted, the treatment time, the depressurisation rate, the temperature and the come-up time (CUT) required to reach the desired pressure should be mentioned [40]. If the CUT is prolonged, it is as if a pretreatment is performed, and the temperature is fundamental, it seems that values of 45–50°C increase the inactivation of pathogens and yeasts [54].

It has been described in various publications that this treatment of HHP, 20–180 Mpa, can produce populations with sublethal damage [30, 68–70]. It is very important to take into account if the treatment carried out in food can produce this type of population since it would produce an estimate of economic life and erroneous security by being able to survive and revive over time even if it is in low concentrations.

The inactivation of *L. monocytogenes* in different meat products has been studied by several authors [60, 71], which reported that pressure treatments of up to 300 MPa are insufficient to inactivate it.

In fermented products such as chorizo, it has been described that the application of HHP can contribute to lowering the altering microbiota, without adverse effects on fermentative bacteria with a treatment of 400 MPa/154 s [42].

The mechanism of action of HHP on microorganisms has been described by various authors that causes damage to the cell membrane [30, 51, 72] and induces morphological changes in the microorganism [73].

The cell membrane is damaged and therefore causes irreversible damage and cell death. It produces crystallisation of the acyl chains of the phospholipid bilayer that leads to bud formation, intracellular material leakage and membrane rupture [30].

Proteins at pressures greater than 100 MPa hydrophobic interactions tend to increase in volume and will cause protein denaturation. In the case of enzymes, it generates conformation changes and, therefore, cell damage and death [34, 74].

There is also inactivation of enzymes related to DNA replication and transcription [34, 74].

2.4 Predictive modelling applied to meat products treated by HPP

Although the effectiveness of HHP application has been recognised by various authors to reduce the levels of various pathogens to acceptable levels in several foods, it is important to take into account the fact that the treatment can be sublethal and only cause lesions in subpopulations of microbial cells. These cells can recover from this type of lesions and grow during the period of storage of the product or before its consumption, reaching levels above the levels allowed by current legislation. Based on this, many authors evaluated and modelled the behaviour of *L. monocytogenes* during and after the treatment of APH in meat products, that is, throughout their useful life [15, 25, 75–77]. These models are essential tools for decision-making in the industry in terms of meeting microbiological criteria. In addition to the predictive models described, there are models in the literature that describe the probability of inactivation/recovery, or also called survival/death (logistic) interface models, of *L. monocytogenes* in meat products or culture media.

Predictive models of inactivation developed in culture media, once validated in specific food matrices such as chorizo, can be applied in the meat industry. Examples of these models would be those developed for *L. monocytogenes* and *L. innocua* (as a surrogated for safety purpose) in meat products [5, 19, 22, 24–26].

In **Table 1**, several types of predictive models that consider treatment inactivation and/or growth on storage phase of meat product can be observed.

Bover-Cid et al. [22] developed and validated a polynomial model of the inactivation of *L. monocytogenes* induced by HPP on dry-cured ham (Eq. (1)), as a function of the technological parameters: pressure intensities (347–852 MPa), pressure holding time (2.3–15.75 min) and fluid temperature (7.6–24.4°C). Pressure and time were the most critical factors influencing microbial inactivation, and the little effect was observed applying pressures below 450 MPa. The increase in holding time for

Reference	Meat product	Equation
Inactivation model during treatment		
Bover-Cid et al. [5]	Cured ham	$\log\left(N/N_0\right) = 38.653 - 34.29 \cdot a_w - 0.0237 \cdot P - 0.00349 \cdot F^2 + 0.000334 \cdot P \cdot F$
Growth model after treatment		
Hereu et al. [14]	Cooked ham	$\log(N) = \log \frac{10^{9.09}}{1+\left(\frac{10^{9.09}}{N_0}-1\right)\cdot \exp\left(-(0.023\cdot(T+1.80))^2\cdot\left(t-\left(\frac{(6.30\cdot 23.85\cdot T+2)\cdot \ln(2)}{(0.023\cdot(T+1.80))^2}\right)\right)\right)}$
Probability of recovery during and after treatment		
Valdramidis et al. [18]	Uncured meat	$\text{Logit}(Pr) = 62.08 - 1.83 \cdot 10^{-1} \cdot P + 1.38 \cdot 10^{-4} \cdot P^2 - 0.18 \cdot 10^{-3} \cdot P \cdot t_s - 4.25 \cdot 10^{-3} \cdot P \cdot a_W$
Koseki and Yanamoto [78]	Saline solution	$\text{Logit}(Pr) = 12.9973 - 0.0775 \cdot P - 9.1909 \cdot \log(t) + 2.3331 \cdot pH + 1.6674 \cdot IC$

N and N_0 represent the final and initial concentrations of the pathogen, respectively.
P = applied pressure; t = treatment time; t_s = storage time; F = fat content; a_w = water activity; T = storage temperature; IC = initial concentration of the pathogen, Pr = probability.

Table 1.
Predictive models obtained during/after the process of inactivation of L. monocytogenes *by HHP on meat products.*

longer than 10 min and the temperature tested did not lead to a significant increase in inactivation of the pathogen.

$$\log\left({}^{N}\!/_{N_0}\right) = -380.3164 + 292.5942 \cdot P_{log} - 56.1268 \cdot {P_{log}}^2 + 1.4090 \cdot t + 0.0133 \cdot t^2 - 0.6423 \cdot P_{log} \cdot t \quad (1)$$

Bover-Cid et al. [5] used the response surface methodology (RSM) (**Table 1**) to evaluate the effect of a_w and fat content in the inactivation of *L. monocytogenes* by HPP in dry-cured ham. Besides these two intrinsic factors, the pressure intensity (347–600 MPa, during 5 min) was also considered as an independent variable for model development. According to the best fitting polynomial equation, all the three factors evaluated influenced on HHP inactivation, reaching inactivation levels from 0.92 to 6.82 logs.

Hereu et al. [25] obtained inactivation curves of *L. monocytogenes* on sliced RTE cooked meat products, ham (Eq. (2)) and mortadella (Eq. (3)) (which differ mainly on fat concentration), during HPP at pressures from 300 to 800 MPa. Their results suggested that the fat content of mortadella would have a protective effect on *L. monocytogenes* to pressure, in comparison with cooked ham. The log-linear with tail primary model was adequate to describe the inactivation kinetics at different holding times, which means that a first-order kinetics was applicable to describe the inactivation before a tailing effect appeared that suggests the presence of a more resistant subpopulation of cells. Secondary model was also performed to establish the relationship between the primary kinetic parameters, log K_{max} and log N_{res}, and pressure treatments. Combining the equations resulted from the primary and secondary modelling approaches; the inactivation of *L. monocytogenes* could be estimated as a function of pressure and holding time

$$\log {}^{N}\!/_{N_0} = \log\left[\left(10^{\log N_0} - 10^{8.0832-0.0121 \cdot P}\right) \cdot e^{-\left(10^{-2.9869+0.0069 \cdot P} \cdot t\right)} + 10^{8.0832-0.0121 \cdot P}\right] - \log(N_0)(\text{cooked ham}) \quad (2)$$

$$\log {}^{N}\!/_{N_0} = \log\left[\left(10^{\log N_0} - 10^{8.6636-0.0125 \cdot P}\right) \cdot e^{-\left(10^{-3.6586+0.0079 \cdot P} \cdot t\right)} + 10^{8.6636-0.0125 \cdot P}\right] - \log(N_0)\ (\text{mortadella}) \quad (3)$$

Hereu et al. [14] built up another model for the estimation of growth of *L. monocytogenes* in sliced cooked meat products (cooked ham and mortadella) after pressurisation but includes other factors as two different inoculum levels (10^7 or 10^4 cfu/g), two physiological states of cells (freeze-stressed or cold-adapted) and different storage temperatures (4, 8, and 12°C). The logistic model with delay (primary model) was fitted to data to estimate the lag phase (λ) and the maximum specific growth rate (μ_{max}). Secondary modelling was performed, using the Ratkowsky square root model (**Table 1**) and the relative lag time (RLT) concept. They observed that the time to achieve a 2-log cfu/g concentration of *L. monocytogenes* was similar for both physiological states. Freeze-stressed cells were more resistant to pressures and showed more extended lag phase during storage than the cold-adapted bacteria.

Based on logistic regression (**Table 1**), [18] concluded that the recovery of *L. monocytogenes* in a simulated cured meat after HPP treatments is influenced by the pressure applied, the storage time and the synergistic effect of pressure and a_w. The effect of salt reduction on the recovery of *L. monocytogenes* following HPP in meat systems was assessed. A protective effect was remarked at low a_w values which led to low inactivation levels both immediately and during storage.

Koseki and Yanamoto [78] developed a probability model (a simple linear logistic regression model, R^2 = 0.9213, **Table 1**) of recovery of *L. monocytogenes* on sliced cooked ham during and after HHP treatment, with a storage of 10°C during 70 days. Authors defined "recovery" as the detection of >10^2 cfu/g bacteria, and the ham score was "1" as when there was a recovery of cells and "0" when not. The treatment applied to 500 MPa for 10 min allowed the reduction of *L. monocytogenes* of 5 logs cfu/g, reaching below the detectable level (10 cfu/g). However, they described a gradual increase of bacterial count during storage that at the end of the experiment, reached 7–8 log cfu/g. This model does not only calculate the appropriate process condition of HPP treatment but also provides information for the estimation of risk of the recovery of *L. monocytogenes* during storage of the product.

Mussa et al. [79] obtained kinetic data on *L. monocytogenes* inactivation by HPP on pork chop samples. The variables studied were pressure intensities (200–400 MPa) and duration of pressure treatments (0–90 min). Interestingly, this is one of the few studies in which the pressure inactivation kinetics was analysed assuming a first-order kinetic process (Eq. (4)):

$$\log\left({}^{N}\!/_{N_0}\right) = -kt \tag{4}$$

where N refers to the number of viable cells in samples after pressure treatments; N_0 is the number of viable cells just before pressures achieved the intensities set in the experimental design; t is the time in minutes; and k is the reaction rate constant (min^{-1}).

The D value, which is the treatment time at any given pressure required to produce one decimal reduction, was calculated as the inverse of the slope (Eq. (5)):

$$D = \left({}^{1}\!/_{slope}\right) \tag{5}$$

Two secondary models were assessed to describe the pressure dependence as a function of kinetic parameters (k and decimal reduction time *D*): Arrhenius-type and the pressure death time (PDT) models. Both models described well the kinetic parameters (R > 0.96). Higher lethal effects were observed when higher pressures were applied, with an increase in K values and a decrease in *D* values as pressure levels increased. The holding time also had a significant effect on inactivation.

The pressure ZHP (the pressure range between which the decimal reduction time changes by a factor of 10) was calculated as the negative of the inverse of the slope of the curve of log D values versus pressure as follows (Eq. (6)):

$$\text{ZHP} = -\left({}^{1}\!/_{slope}\right) \tag{6}$$

Results indicated that to achieve a 5 log cfu/g reduction of *L. monocytogenes* levels, approximately 7.5 min of pressure holding time, when pressure is set to be 400 MPa (D value = 1.49 min), would be necessary. At the same conditions of this study, Mussa et al. [79] obtained a D value = 3.52 min on pork, which makes clear that, besides the technological parameters, the type and composition of food influence on the destruction kinetics of *L. monocytogenes* by HPP.

Oliveira et al. [80] evaluated the effect of HPP (600 MPa/180 s at 25°C) in combination with the application of natural phenolic bioactive carvacrol (at 200 ppm) to reduce *Listeria innocua* levels in a low-sodium sliced vacuum-packed turkey breast ham during 60 days of storage at 4°C. The initial contamination of slices with *L. innocua* was ~10^6 cfu/g of slice. The primary model of Baranyi and Roberts, fitted to data obtained during the storage period, showed a significant extension of shelf life of low-sodium vacuum-packed turkey breast ham, with the reduction of maximum population density and the increase in lag phase duration. *L. innocua* has been used as a surrogate of *L. monocytogenes* for processing plant safety purposes, as it has similar physiological and metabolic characteristics to those of pathogenic species [81].

The effect of HPP treatments and potassium lactate on inactivation of *L. monocytogenes* was evaluated by Lerasle [26] considering the variables pressure intensities (200–500), holding time (2–14 min) and potassium lactate concentrations of 0 or 1.8% w/w. The Weibull model was fitted to the inactivation data (*log N* versus *time*) obtained at the different pressure holding times evaluated. The secondary model was a linear regression that defines log b as a function of the pressure intensity and explanatory factors (Eq. (7)). Considering that the lactate concentration effect was not significant (ANOVA, $p > 0.05$), the secondary model was:

$$\log b = \log b * - \frac{P}{Z_p} + \varepsilon; \tag{7}$$

where Z_p might be interpreted as the pressure required to reduce b by 10-fold and $logb^*$ is the y-intercept and ε the model error. The estimated values for the parameters are represented below (Eq. (8)):

$$\log b = 143 - \frac{P}{3.1} + \varepsilon \tag{8}$$

Combining primary and secondary models makes it possible to recalculate the log reduction obtained at various times and pressures intensities.

These models developed by Lerasle et al. [26] were subsequently applied in a multi-criteria framework combining safety, hygiene and sensorial quality to investigate the possibility of extending the shelf life of a ready-to-cook poultry product, using the HPP technology [82]. Models developed for *Salmonella* and *E. coli* were also considered in the framework in which the maximum allowed contamination level of *L. monocytogenes* was set to be 100 cfu/g (according to the microbiological criteria of the foodstuffs defined by the Commission Regulation (EC) No 2073/2005) [1]. The approach is a decision support tool for shelf life determination.

Also the significant inactivation effect ($P < 0.001$) of HHP (540 MPa/270 s) on *Enterobacteriaceae*, *E. coli* and *Pseudomonas*, coagulase-negative *Staphylococcus* (CNS) and LAB of natural casings and condiments used in the processing of cured meat sausage using response surface methodology was described by Fraqueza et al. [83]. Treated casings turned slightly whiter, but their resistance (FT) to breakage (i.e. casings structural integrity) was not affected.

Recently, Guillou and Membré [52] have carried out a hierarchical model based on a study of metadata of the determining factors in inactivation by the treatment of high pressures in different microorganisms and substrates, concluding that those more relevant factors studied were the species, the strain and the pH and that the most resistant species was *Staphylococcus* and the most sensitive *Salmonella*.

Novel approaches have been described as the potential use of Listex™ P100 in sausage “Alheira” combined with high hydrostatic pressure, applying Weibull model [84] and concluding that at mild HHP treatment, phage P100 remained active and seemed to present potential to be added in nonthermal inactivation of *L. monocytogenes*.

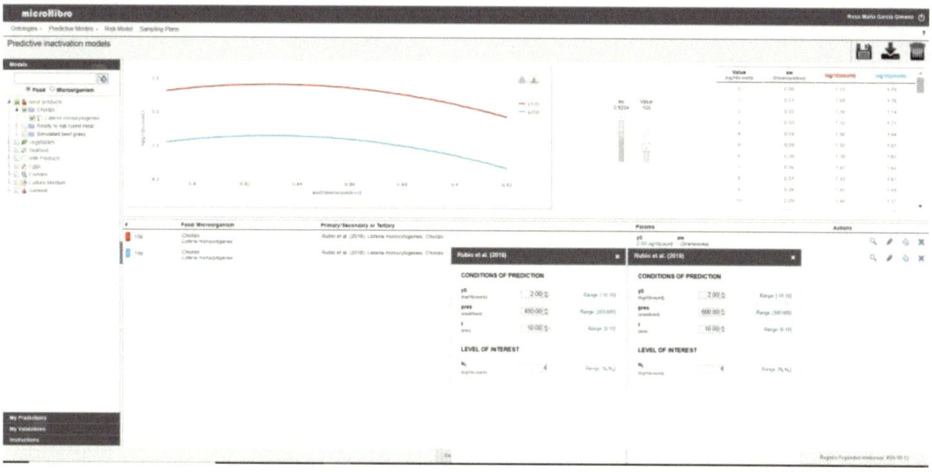

Figure 1.
Screenshot of the microHibro web application of a predictive model of HHP treatment of chorizo.

High pressure processing and biopreservation can contribute to food safety by inactivation of bacterial contaminants. However, these treatments are inefficient against bacterial endospores such as *Bacillus* and *Clostridium* species. Moreover, HPP can induce spore germination [85]. In [85], it is reported that *Lactococcus lactis* strain CH-CH15 was able to regrow after HPP treatments, thus an excellent option to be preservative against *Bacillus* and *Clostridium* strains during chilled storage. The inactivation model used was fitted by using a reparametrized Weibull model, whereas growth curves of lactic acid bacteria were modelled with a logistic model.

Predictive microbiology modelling easy-to-use software has been developed to allow users involved in food safety management to use a tool to asses them and help them for decision-making. Several applications have been developed, but just a few had incorporated the prediction of HHP treatment. One of it is the "HP3", available online (www.hp3.cat) elaborated by the Institute of Agrifood Research and Technology (Spain), and another is microHibro (www.microhibro.com), built up by the University of Córdoba (Spain) (**Figure 1**).

For further information, there are several reviews as [4, 10, 23, 30, 31, 78].

2.5 Other applications of HPP on meat products

Although the main application of HHP is enzymatic and microbial inactivation to extend commercial life and inactivate pathogens, other possible applications such as obtaining different types of fish, meat, egg and milk gels have been described. Likewise, this technology accelerates the diffusion of solutes in various foods, the solubilisation of gases and the extraction processes. The possibility of using high pressures to keep food at temperatures below 0°C in a liquid state (at 207.5 MPa, the water remains liquid at temperatures of −22°C) or to induce freezing (supercooling) and ultra-fast defrosting constitutes a promising new field of study and application in the food industry [34, 40]. Also applying low pressures, 100–150 MPa have been employed to tenderised pre-rigour meat of rabbit, chicken, pork and beef. Higher pressures, 250 MPa, has been applied, for example, before smoking to treat roast beef and bacon, to inactivate microflora of minced meat or to treat foie gras to extend shelf life [40, 41].

3. Conclusion

High hydrostatic pressure treatment has been described to improve the microbiological safety and shelf life of ready-to-eat meat products, as a nonthermal decontamination technology in the meat industry, applied at pre- or post-packaging. There are a variety of factors that influence the treatment effect that should be taken into account when applied to food. The pathogen widely studied in this product is *L. monocytogenes* that reflects the concern of the food industry. The predictive modelling is an important tool of the novel microbial food safety management strategy that provides with accurate information to demonstrate and guarantee the safety and shelf life of the food products and also helps to the optimisation of the meat treatment.

Acknowledgements

We greatly acknowledged the Spanish government and European ERDF funding for supporting the RTA-2013-00055-C02-02 project, providing material and especially human resources, making possible the continuation of risk assessment and management activities at national and European level by our research group AGR170 "HIBRO".

Author details

Rosa María García-Gimeno* and Guiomar Denisse Posada Izquierdo
Department of Food Science and Technology, Agrifood Campus of International Excellence, Universidad de Córdoba, Córdoba, Spain

*Address all correspondence to: bt1gagir@uco.es

References

[1] Commission E. Commission regulation (EC) N0. 2073/2005 of 15 November 2005 on microbiological criteria for foodstuffs. Official Journal of the European Communities. 2005;**338**: 1-26

[2] European Food Safety Authority and European Centre for Disease Prevention and Control (EFSA and ECDC). The European Union summary report on trends and sources of zoonoses, zoonotic agents and food-borne outbreaks in 2017. EFSA Journal. 2018;**16**(12):5500

[3] Nastasijevic I, Milanov D, Velebit B, Djordjevic V, Swift C, Painset A, et al. Tracking of *Listeria monocytogenes* in meat establishment using whole genome sequencing as a food safety management tool: A proof of concept. International Journal of Food Microbiology. 2017;**18**:157-164

[4] Buzrul S. Multi-pulsed high hydrostatic pressure treatment of foods. Food. 2015;**4**:173-183

[5] Bover-Cid S, Belletti N, Aymerich T, Garriga M. Modeling the protective effect of aw and fat content on the high pressure resistance of *Listeria monocytogenes* in dry-cured ham. Food Research International. 2015;**75**:194-199

[6] Hayagreeva D, Pandey M. Novel approaches in improving the quality and safety aspects of processed meat products through high pressure processing technology - a review. Trends in Food Science and Technology. 2016;**54**:175-185

[7] Guyon C, Meynier A, Lamballerie M. Protein and lipid oxidation in meat: A review with emphasis on high-pressure treatments. Trends in Food Science & Technology. 2016;**50**:131-143

[8] Hugas M, Garriga M, Monfort J. New mild technologies in meat processing: High pressure as a model technology. Meat Science. 2002;**62**(3):359-371

[9] Garriga M, Grèbol N, Aymerich MT, Monfort JM, Hugas M. Microbial inactivation after high-pressure processing at 600 MPa in commercial meat products over its shelf life. Innovative Food Science & Emerging Technologies. 2004;**5**(4):451-457

[10] Rendueles E, Omer MK, Alvseike O, Alonso-Calleja C, Capita R, Prieto M. Microbiological food safety assessment of high hydrostatic pressure processing: A review. LWT- Food Science and Technology. 2011;**44**(5):1251-1260

[11] Bajovic B, Bolumar T, Heinz V. Quality considerations with high pressure processing of fresh and value added meat products. Meat Science. 2012;**92**(3):280-289

[12] Ananou S, Garriga M, Jofré A, Aymerich T, Gálvez A, Maqueda M, et al. Combined effect of enterocin AS-48 and high hydrostatic pressure to control food-borne pathogens inoculated in low acid fermented sausages. Meat Science. 2010;**84**: 594-600

[13] Balamurugan S, Ahmed R, Chibeu A, Gao A, Koutchma T, Strange P. Effect of salt types and concentrations on the high-pressure inactivation of *Listeria monocytogenes* in chapter II 77 ground chicken. International Journal of Food Microbiology. 2016;**218**:51-56

[14] Hereu A, Dalgaard P, Garriga M, Aymerich T, Bover-Cid S. Analysing and modelling the growth behaviour of *Listeria monocytogenes* on RTE cooked meat products after a high pressure treatment at 400 MPa. International Journal of Food Microbiology. 2014;**186**: 84-94

[15] Hereu A, Bover-Cid S, Garriga M, Aymerich T. High hydrostatic pressure and biopreservation of dry-cured ham to meet the food safety objectives for *Listeria monocytogenes*. International Journal of Food Microbiology. 2012a;**154** (3):107-112

[16] Patterson M, Mackle A, Linton M. Effect of high pressure, in combination with antilisterial agents, on the growth of *Listeria monocytogenes* during extended storage of cooked chicken. Food Microbiology. 2011;**28**(8): 1505-1508

[17] Porto-Fett ACS, Call JE, Shoyer BE, Hill DE, Pshebniski C, Cocoma GJ, et al. Evaluation of fermentation, drying, and/or high pressure processing on viability of *Listeria monocytogenes, Escherichia coli* O157:H7, *Salmonella* spp., and *Trichinella spiralis* in raw pork and Genoa salami. International Journal of Food Microbiology. 2010;**140**: 61-75

[18] Valdramidis VP, Patterson MF, Linton M. Modelling the recovery of *Listeria monocytogenes* in high pressure processed simulated cured meat. Food Control. 2015;**47**:353-358

[19] Rubio B, Possas A, Rincón F, García-Gimeno RM, Martínez B. Model for *Listeria monocytogenes* inactivation by high hydrostatic pressure processing in Spanish chorizo sausage. Food Microbiology. 2018;**69**:18-24

[20] Hjelmqwist J. Commercial high pressure equipment. In: Barbosa-Cánovas GV, Tapia MS, Cano MP, editors. Novel Food Processing Technologies. New York: CRC Press; 2005. pp. 361-374. EEUU

[21] Possas A, Valdramidis V, García-Gimeno RM, Pérez-Rodríguez F. High hydrostatic pressure processing of sliced fermented sausages: A quantitative exposure assessment for *Listeria monocytogenes*. Innovative Food Science and Emerging Technologies. 2019;**52**: 406-419

[22] Bover-Cid S, Belletti N, Garriga M, Aymerich T. Model for listeria monocytogenes inactivation on dry-cured ham by high hydrostatic pressure processing. Food Microbiology. 2011; **28**(4):804-809

[23] Possas A, Pérez-Rodríguez F, Valero A, García-Gimeno RM. Modelling the inactivation of *Listeria monocytogenes* by high hydrostatic pressure processing in foods: A review. Trends in Food Science & Technology. 2017;**70**:45-55

[24] Carlez A, Rosec J, Richard N, Cheftel J. High pressure inactivation of *Citrobacter freundii, Pseudomonas fluorescens* and *Listeria innocua* in inoculated minced beef muscle. LWT - Food Science and Technology. 1993; **26**(4):357-363

[25] Hereu A, Dalgaard P, Garriga M, Aymerich T, Bover-Cid S. Modeling the high pressure inactivation kinetics of *Listeria monocytogenes* on RTE cooked meat products. Innovative Food Science and Emerging Technologies. 2012b;**16**: 305-315

[26] Lerasle M, Guillou S, Simonin H, Anthoine V, Chealret R, Federighi M, et al. Assessment of *Salmonella* and *Listeria monocytogenes* level in ready-to-cook poultry meat: Effect of various high pressure treatments and potassium lactate concentrations. International Journal of Food Microbiology. 2014;**186**: 74-83

[27] AESAN (Agencia Española de Seguridad Alimentaria y Nutrición). Opinión del Comité científico de la AESA sobre una cuestión presentada por la Dirección Ejecutiva, en relación con la aplicación de altas presiones en carne y productos cárnicos (Ref.: AESA-2003-007) Revista del Comité Científico de la AESAN, 1; 36–71; 2005

[28] Codex Alimentarius. Guidelines on the application of general principles of food hygiene to the control of *Listeria monocytogenes* in foods. Codex Alimentarius Commission, CAC/GL. 2007;**61**:1-28

[29] USDA/FSIS. Compliance Guidelines to Control *Listeria monocytogenes* in Postlethality Exposed Ready-to-Eat Meat and Poultry Products. US Department of Agriculture: Washington D.C; 2006. Available from: http://www.fsis.usda.gov/oppde/rdad/FRPubs/97-013F/LM_Rule_Compliance_Guidelines_May_2006.pdf

[30] Serment-Moreno V, Barbosa-Cánovas G, Torres J, Welti-Chanes J. High-pressure processing: Kinetic models for microbial and enzyme inactivation. Food Engineering Reviews. 2014;**6**(3):56-88

[31] Campus M. High pressure processing of meat, meat products and seafood. Food Engineering Reviews. 2010;**2**:256-273

[32] Tonello C. Case studies on high pressure processing of foods. In: Zhang HQ, BarbosaCánovas GV, Balasubramaniam VM, Dunne CP, Farkas J, Yuan JTC, editors. Nonthermal Processing Technologies for Food. Oxford: Wiley-Blackwell; 2011. pp. 36-50

[33] Das S, Lalitha K, Joseph G, Kamalakanth C, Bindu J. High pressure destruction kinetics along with combined effect of potassium sorbate and high pressure against *Listeria monocytogenes* in Indian white prawn muscle. Annals of Microbiology. 2016; **66**(1):245-251

[34] Pina Pérez MC. Aplicación de tecnologías no térmicas de conservación, pulsos eléctricos de alta intensidad (PEAI) y altas presiones hidrostáticas (APH), para el control de *Cronobacter sakazakii* en fórmulas lácteas infantiles: desarrollo de modelos predictivos y evaluación de la exposición [tesis doctoral]. Editorial Universitat Politècnica de València; 2011

[35] Barbosa-Canovas GV, Ibarz A. Introduction to Food Process Engineering. New York: CRC Press; 2014. EEUU

[36] Morales P, Calzada J, Ávila M, Nuñez M. Inactivation of *Escherichia coli* O157: H7 in ground beef by single-cycle and multiple-cycle high-pressure treatments. Journal of Food Protection. 2008;**71**(4):811-815

[37] Morales P, Calzada J, Rodríguez B, de Paz M, Nuñez M. Inactivation of *Salmonella Enteritidis* in chicken breast fillets by single-cycle and multiple-cycle high pressure treatments. Foodborne Pathogens and Disease. 2009;**6**(5): 577-581

[38] Patazca E, Koutchma T, Balasubramaniam V. Quasiadiabatic temperature increase during high pressure processing of selected foods. Journal of Food Engineering. 2007; **80**(1):199-205

[39] McArdle R, Marcos B, Kerry J, Mullen A. Monitoring the effects of high pressure processing and temperature on selected beef quality attributes. Meat Science. 2010;**86**(3):629-634

[40] Welti-Chanes J, López-Malo E, Palou D, Bermúdez J, Guerrero-Beltrán G, Barbosa-Cánovas G. Fundamentals and applications of high pressure processing to foods. In: Barbosa-Cánovas GV, Tapia S, Cano MP, editors. Novel Food Processing Technologies. New York: Marcel Dekker. EEUU; 2005

[41] Cheftel JC. Effects of high hydrostatic pressure on food constituents: An overview. In: Balny C, Hayashi R, Heremans K, Masson P, editors. High Pressure and

Biotechnology, Colloque INSERM, Vol. 224. Montrouge, France: John Libbey Eurotext; 1992. pp. 195-209

[42] Alfaia A, Alfaia C, Patarata L, Fernandes MJ, Fernandes MH, Elias M, et al. Binomial effects of high isostatic pressure and time on the microbiological, sensory characteristics and lipid composition stability of vacuum packed dry fermented sausages “chouriço”. Innovative Food Science and Emerging Technologies. 2015;**32**: 37-44

[43] Black E, Huppertz T, Fitzgerald G, Kelly A. Baroprotection of vegetative bacteria by milk constituents: A study of *Listeria innocua*. International Dairy Journal. 2007;**17**:104-110

[44] Gao Y, Ju X, Wu D. A predictive model for the influence of food components on survival of *Listeria monocytogenes* LM 54004 under high hydrostatic pressure and mild heat conditions. International Journal of Food Microbiology. 2007;**117**(3): 287-294

[45] Garcia-Graells C, Masschalck B, Michiels C. Inactivation of *Escherichia coli* in Milk by high-hydrostatic pressure treatment in combination with antimicrobial peptides. Journal of Food Protection. 1999;**62**(11):1248-1254

[46] Kaur L, Astruc T, Vénien A, Loison O, Cui J, Irastorza M, et al. High pressure processing of meat: Effects on ultrastructure and protein digestibility. Food & Function. 2016;**7**(5):2389-2397

[47] Huang Y, Wang Y, Wu Z, Li F. Combined effects of high-pressure and thermal treatments on lipid oxidation and enzymes in pork. Food Science and Biotechnology. 2016;**25**(1):261-266

[48] Ferrini G, Comaposada J, Arnau J, Gou P. Colour modification in a cured meat model dried by quick-dry-slice process® and high pressure processed as a function of NaCl, KCl, K-lactate and water contents. Innovative Food Science & Emerging Technologies. 2012;**13**: 69-74

[49] Ros-Polski V, Koutchma T, Xue J, Defelice C, Balamurugan S. Effects of high hydrostatic pressure processing parameters and NaCl concentration on the physical properties, texture and quality of white chicken meat. Innovative Food Science & Emerging Technologies. 2015;**30**:31-42

[50] Siddig H, Salah A, Aljasass F. Effect of high hydrostatic pressure treatment on the microbial contamination and on some chemical and physical properties of minced chicken. Research Journal of Biotechnology. 2019;**14**(9): 89-95

[51] Patterson M. Microbiology of pressure-treated foods. Journal of Applied Microbiology. 2005;**98**(6): 1400-1409

[52] Guillou S, Membré J. Inactivation of *Listeria monocytogenes, Staphylococcus aureus*, and *Salmonella enterica* under high hydrostatic pressure: A quantitative analysis of existing literature data. Journal of Food Protection. 2019;**82**(10):1802-1814

[53] Barbosa-Cánovas G, Gongora-Nieto M, Pothakamury U, Swanson B. Preservation of Foods with Pulsed Electric Fields. San Diego: Academic Press. EEUU; 1999

[54] Zook CD, Parish ME, Braddock RJ, Balaban MO. High pressure inactivation kinetics of *Saccharomyces cerevisiae* ascospores in orange and apple juice. Journal of Food Science. 1999;**64**: 533-535

[55] Oliveira T, Haddad G, Ramos A, Ramos E, Piccoli R, Cristianini M. Optimization of high pressure processing to reduce the safety risk of low-salt ready-to-eat sliced Turkey

breast supplemented with carvacrol. British Food Journal. 2019;**121**(11): 2592-2606

[56] Rodríguez-Calleja JM, Cruz-Romero MC, Sullivan MGO, García-López ML, Kerry JP. High-pressure-based hurdle strategy to extend the shelf-life of fresh chicken breast fillets. Food Control. 2012;**25**:516-524

[57] MacFarlane J, McKenzie I, Turner R. Pressure induced pH and length changes in meat. Meat Science. 1980;**7**:169-181

[58] Georget E, Sevenich R, Reineke K, Mathys A, Heinz V, Callanan M. Inactivation of microorganisms by high isostatic pressure processing in complex matrices: A review. Innovative Food Science & Emerging Technologies. 2015;**27**:1-14

[59] Hayman M, Kouassi G, Anantheswaran R, Floros J, Knabel S. Effect of water activity on inactivation of *Listeria monocytogenes* and lactate dehydrogenase during high pressure processing. International Journal of Food Microbiology. 2008;**124**(1):21-26

[60] Jofré A, Aymerich T, Grèbol N, Garriga M. Efficiency of high hydrostatic pressure at 600 MPa against food-borne microorganisms by challenge tests on convenience meat products. LWT- Food Science and Technology. 2009;**42**(5):924-928

[61] Gayán E, Torres JA, Paredes-Sabja D. Hurdle approach to increase the microbial inactivation by high pressure processing: Effect of essential oils. Food Engineering Reviews. 2012;**4**(3):141-148

[62] Chien SY, Sheen S, Sommers C, Sheen LY. Combination effect of high-pressure processing and essential oil (*Melissa officinalis* extracts) or their constituents for the inactivation of *Escherichia coli* in ground beef. Food and Bioprocess Technology. 2019;**12**(3): 359-370

[63] Park S, Sohn K, Shin J, Lee H. High hydrostatic pressure inactivation of *Lactobacillus viridescens* and its effects on ultrastructure of cells. International of Journal Food Science and Technology. 2001;**36**:775-781

[64] Molina-Hoppner A, Doster W, Vogel R, Ganzle M. Protective effect of sucrose and sodium chloride for *Lactococcus lactis* during sublethal and lethal high-pressure treatments. Applied and Environmental Microbiology. 2004; **70**:2013-2020

[65] Hauben K, Bernaerts K, Michiels C. Protective effect of calcium on inactivation of *Escherichia coli* by high hydrostatic pressure. Journal of Applied Microbiology. 1998;**85**:678-684

[66] Baert L, Debevere J, Uyttendaele M. The efficacy of preservation methods to inactivate foodborne viruses. International Journal of Food Microbiology. 2009;**131**(2-3):83-94

[67] Calci K, Meade G, Tezloff R, Kingsley D. High-pressure inactivation of hepatitis a virus within oysters. Applied and Environmental Microbiology. 2005;**71**(1):339-343

[68] Pagan R, Mackey B. Relationship between membrane damage and cell death in pressure-treated *Escherichia coli* cells: Differences between exponential- and stationary-phase cells and variation among strains. Applied and Environmental Microbiology. 2000;**66**: 2829-2834

[69] Pagan R, Jordan S, Benito A, Mackey B. Enhanced acid sensitivity of pressure-damaged *Escherichia coli* O157:H7 cells. Applied and Environmental Microbiology. 2001;**67**: 1983-1985

[70] Wouters PC, Glaasker E, Smelt JPPM. Effects of high pressure on inactivation kinetics and events related to proton in *Lactobacillus plantarum*.

Applied and Environmental Microbiology. 1998;**64**:509-514

[71] Aymerich T, Jofré A, Garriga M, Hugas M. Inhibition of *Listeria monocytogenes* and *Salmonella* by natural antimicrobials and high hydrostatic pressure in sliced cooked ham. Journal of Food Protection. 2005;**68**:173-177

[72] Patterson MF, Linton M. Pasteurización de alimentos por altas presiones. In: Instituto Tomás Pascual Sanz para la Nutrición y la Salud; Universidad de Burgos (Eds.) Nuevas tecnologías en la conservación y transformación de los alimentos. International Marketing and Communication. España; 2010. pp. 59-72

[73] Ferreira M, Almeida A, Delgadillo I, Saraiva J, Cunha Â. Susceptibility of *Listeria monocytogenes* to high pressure processing: A review. Food Reviews International. 2016;**32**:377-399

[74] Téllez-Luis S, Ramírez J, Pérez-Lamela C, Vázquez M, Simal-Gándara J. Aplicación de la alta presión hidrostática en la conservación de los alimentos. Ciencia y Tecnologia Alimentaria. 2001; **3**(2):66-80

[75] Jofré A, Aymerich T, Bover-Cid S, Barriga M. Inactivation and recovery of *Listeria monocytogenes, Salmonella enterica* and *Staphylococcus aureus* after high hydrostatic pressure treatments up to 900 MPa. International Microbiology. 2010;**13**:105-112

[76] Stollewerk K, Jofré A, Comaposada J, Arnau J, Garriga M. The effect of NaCl-free processing and high pressure on the fate of *Listeria monocytogenes* and *Salmonella* on sliced smoked dry-cured ham. Meat Science. 2012;**90**(2):472-477

[77] Stollewerk K, Jofré A, Comaposada J, Arnau J, Garriga M. NaCl-free processing, acidification, smoking and high pressure: Effects on growth of *Listeria monocytogenes* and *Salmonella enterica* in QDS processed® dry-cured ham. Food Control. 2014;**35**(1):56-64

[78] Koseki S, Yamamoto K. Modelling the bacterial survival/death interface induced by high pressure processing. International Journal of Food Microbiology. 2007;**116**(1):136-143

[79] Mussa D, Ramaswamy H, Smith J. High-pressure destruction kinetics of *Listeria monocytogenes* on pork. Journal of Food Protection. 1999;**62**(1):40-45

[80] Oliveira T, Junior B, Ramos A, Ramos E, Piccoli R, Cristianini M. Phenolic carvacrol as a natural additive to improve the preservative effects of high pressure processing of low-sodium sliced vacuum-packed Turkey breast ham. LWT- Food Science and Technology. 2015;**64**(2):1297-1308

[81] Saucedo-Reyes D, Marco-Celdrán A, Pina-Pérez MC, Rodrigo D, Martínez-López A. Modeling survival of high hydrostatic pressure treated stationary- and exponential-phase *Listeria innocua* cells. Innovative Food Science and Emerging Technologies. 2009;**10**(2): 135-141

[82] Guillou S, Lerasle M, Simonin H, Federighi M. High pressure processing of meat and meat products. In: Cummins EJ, Lyng J, editors. Emerging Technologies in Meat Processing. 1st ed. New York: John Wiley & Sons, Ltd.; 2016. pp. 35-99

[83] Fraqueza M, Martins C, Gama L, Fernandes MH, Fernandes MJ, Ribeiro M, et al. High hydrostatic pressure and time effects on hygienic and physical characteristics of natural casings and condiments used in the processing of cured meat sausage. Innovative Food Science and Emerging Technologies. 2019;**58**:102242

[84] Komora N, Bruschi C, Ferreira V, Maciel C, Brandão T, Fernandes R, et al.

The protective effect of food matrices on *Listeria* lytic bacteriophage P100 application towards high pressure processing. Food Microbiology. 2018;**76**: 416-425

[85] Ramaroson M, Guillou S, Rossero A, Rezé S, Anthoine V, Moriceau N, et al. Selection procedure of bioprotective cultures for their combined use with high pressure processing to control spore-forming bacteria in cooked ham. International Journal of Food Microbiology. 2018;**276**:28-38

Chapter 8

Gamma Irradiation and High Hydrostatic Pressure Applied to Hamburger Conservation

Michelle Guimarães Horta, Fabiana Regina Lima, Carlos Alberto Gois Suzart and Poliana Mendes De Souza

Abstract

Human nutrition is an essential process, since it provides the essential nutrients for their development. Animal source foods are rich in protein, amino acids, vitamins, and minerals. And they are subject to contaminants from the raw material to the final consumption. To avoid microbial contamination and deterioration, various technologies are used to ensure their innocuity. These include gamma irradiation and high hydrostatic pressure (HHP), which are nonthermal treatments. Such treatments may reduce the known adverse effects that occur during thermal processing. In meat products, these technologies may induce lipid oxidation, and to limit this process, the addition of synthetic or natural food antioxidants or both are used. This chapter discusses the use of gamma irradiation, high hydrostatic pressure, and application of natural antioxidants in beef hamburger to ensure their quality.

Keywords: hamburger, gamma irradiation, high hydrostatic pressure, natural food additives, nonthermal technologies

1. Introduction

The nutrition is an essential process for humans as they provide the essential nutrients. Meat and meat products have a prominent position among foods in the human diet, since it is rich in high-quality protein, essential amino acids, B vitamins, minerals, and other nutrients [1, 2] . These nutrients are important in the formation of enzymes, hormones, antibodies, structural proteins, and transporters as well as in the construction and maintenance of tissues [3].

Meats are defined as muscle tissues, without or not include their bone base, and can come from different animal species as long as they are fit for consumption. Meat products are those obtained from meat, edible parts of different animal species in which the properties of the raw materials are modified by means of physical, chemical, or biological treatment techniques or a combination of these methods [4]. These techniques in general may involve the addition of ingredients or co-adjuvants of technology for the production of industrialized meat.

The nutritional composition of meat and meat products favors the possibility microorganism's proliferation [5]. Contamination of meat products may occur during processing and handling of food such as slaughtering, deboning, cutting, fragmentation, processing, packaging, storage, and so forth. This may occur due to extrinsic and/or intrinsic factors such as water, air, soil, temperature, and pH [6, 7].

To improve food safety and quality, various technologies have been used and developed to preserve and protect food against microbial contamination and deterioration. These technologies include nonthermal, thermal, biological, and chemical treatments. Thermal treatments are efficient in inactivation of microorganisms but have the disadvantage of generating unwanted biochemical reactions, since temperature-altered treatment favors changes in food quality. Texture, color, vitamin amounts, and development of unpleasant flavors are included in this change [8].

As an alternative to thermal treatments, the nonthermal treatments were developed. Among them, we can list gamma, electron, X-ray irradiation, high hydrostatic pressure (HHP), and the addition of natural antimicrobials [5]. These processes do not use temperature as a way to inactivate microorganisms and enzymes [8], and further generally, the nonthermal treatments do not affect their nutritional and sensory characteristics [8]. The use of nonthermal treatments in meats and derivatives for industrial productions is shown in several studies that among them, the irradiation and HHP are the ones that offer practical possibilities of application [9].

The production of meat industrialized is a strategy for total or partial use of less noble meat. In this class of derivates, sausages, cured meat, ham, hamburger, meatball, and others are included.

The hamburger originated in the city of Hamburg, located in Germany, and this product was consumed raw. In the 1920s, it emerged in the United States. In Brazil, it arrived in the 1950s and became known after it was produced and distributed by fast food chain [10]. It is defined as a meat product, obtained from ground beef of different animal species, with or without the addition of ingredients, molded as a disc or an oval, and subjected to a specific technological process [4, 11]. Also, according to the US Federal Code of Regulation [12], the hamburger is defined as "fresh or frozen ground beef steak, with or without added fat and/or condiments, which should not contain more than 30% fat and should not contain added water."

In relation to world beef production, the United States produces about 19%, followed by Brazil with 17%, the European Union and China with 13% each, and India with 7% [13]. Brazil exports approximately US$ 500 million/month, being considered the largest exporter of beef [14]. In the year 2018, from January to September, 23.47 million heads of cattle were slaughtered [15]. Research conducted by the US Department of Agriculture is estimated that in 2019, there will be a 3% increase in production and a 5% increase in beef exports in Brazil [16].

The quality, safety, and nutritional profile of beef depend on several factors, such as genetic characteristics and animal feeding, slaughter, processing, handling, and others [13].

This chapter discusses the use of emerging nonthermal technologies in meats and derivatives for industrial production, in highlighting among them the irradiation and HHP, to ensure their sage consumption.

2. Nonthermal treatments

A discussion of the most representative nonthermal treatments is shown in this section. Further to the application of natural antioxidants among nonthermal treatments, irradiation and HHP are the ones that offer practical application possibilities in meats and derivatives for industrial production [9]. Despite the efficiency of these nonthermal treatments in food conservation, they may favor lipid oxidation of meat products. To avoid this, there is a tendency to combine these treatments with the use of natural antioxidants to reduce the sensory changes that are caused by oxidation [17].

2.1 Gamma irradiation

Irradiation is a physical treatment in which the food is exposed to a defined ionizing radiation dose. The purpose of this treatment is to control insect infestation, reduce the number of pathogenic microorganisms or deterioration, delay, or eliminate natural biological processes (ripening, germination, or sprouting in fresh foods) [18].

According to [19], irradiated foods were evaluated by several surveys and tests over several years, thus ensuring a safe food for consumption, in relation to nutritional adequacy and toxicological and microbiological safety.

Before 1997, the use of irradiation was limited. However, after Food and Drug Administration (FDA) was approved of the use of irradiation in refrigerated or frozen meats and derivatives to control food-borne pathogens, the consumers began to check the benefits of irradiated food [20, 21].

Irradiation can be applied to any kind food. The Codex Alimentarius Commission [22] regulates that the maximum safe dosage for food in general is 10 kGy, where the minimum dosage absorbed by the food must be sufficient to achieve the technological purpose and the maximum dosage absorbed must not compromise the consumer's health or cause the food to be disposed of [22].

According to Codex Alimentarius Commission [22], there is no minimum dosage to be used in meats, but the corresponding maximum dosages of 4.5 kGy for refrigerated beefs, 7 kGy for frozen meats, and 3 kGy for poultry meats have been defined. In Brazilian legislation [23], it is defined for any food that the minimum dose should be sufficient to achieve the purpose, and the maximum should be lower than that which would compromise the functional and sensorial characteristics of the food.

According to Codex Alimentarius Commission [22] for irradiated foods, the radiation sources that can be used are gamma rays, the radionuclides ^{60}Co, or ^{137}Cs; X-rays generated with a maximum level of 5 MeV; and electrons generated with a maximum level of 10 MeV. These sources have high energy that changes the position of the electrons of the atoms and molecules, converting them into electrically charged particles (ions). It should be noted that these energies are not capable of inducing radioactivity in any material [18].

Irradiation is considered one of the best emerging technologies to guarantee microbiological safety, in which any food can be irradiated, including meats and derivatives [18]. It is effective in eliminating or reducing pathogenic microorganisms, such as *Listeria monocytogenes*, *Salmonella* spp., *Escherichia coli*, and *Staphylococcus aureus* [24].

Although the numerous advantages of the use of the irradiation in the foods, the use this process in the meat products may favor physicochemical and biochemical alterations, as for example the lipid oxidation increase. This oxidation is characterized by the formation of free radicals, that is, it is initiated in the unsaturated fraction of fatty acids by the uptake of a hydrogen atom and propagated as a radical-mediated chain reaction. This process depends on the chemical composition of the meat and access to light, oxygen, and storage temperature. As it causes the increase of oxidation, there are changes in taste, aroma, and nutritional value changes that affect the quality of the food [25–27]. The addition of antioxidants may contribute to the reduction of this process [28].

Regarding the label of the irradiated food, according to the Codex, food that goes through the irradiation process must include on the label a statement indicating that the treatment has taken place and may optionally use the international symbol available [29]. However, under US law, the symbol must be used and must be declared: "Treated with radiation" or "Treated by irradiation." The Brazilian regulation should include in the main panel the words "FOOD TREATED BY IRRADIATION PROCESS" [23].

Table 1 shows data from the study by Kume et al. [30] that reported the quantity of meat and fish products that were irradiated in the listed countries.

In a research conducted by Chirinos et al. [31], samples of industrialized hamburger inoculated with *Escherichia coli* O157: H7 were subjected to the irradiation process. It was found that at low doses (1.08 kGy), it was sufficient to reduce the microorganism, without rejection by the trained tasters.

It was evaluated by Moura [32], the oxidation of cholesterol in beef burgers and chicken burgers submitted to irradiation and stored under freezing conditions. It was found that there was an 11% increase in cholesterol oxide levels in frozen burgers.

Frozen chicken hamburger was inoculated with *Salmonella* sp. and irradiated in a study conducted by Vieira [33], and it was found that dosages of 5 and 7 kGy would be sufficient to reduce the population of *Salmonella* sp. Sensory evaluation did not change significantly, and shelf life was 120 days the same as the conventional product.

2.2 High hydrostatic pressure

The effect of high hydrostatic pressure (HHP) processing was first reported in 1899 by Hite [34]. This process uses an isostatic pressure at room temperature between 100 and 600 MPa. The pressure in the closed and degassed chamber is transmitted by pumps through a liquid (usually uses water) uniformly and instantaneously, which causes the molecular volume to change. The physical effect of the process occurs in the molecules, in which bonds that are weaker, such as those of hydrogen and hydrophobic, are modified [9].

The behavior of food under pressure is determined by three principles: Le Chatelier principle (any reaction that is accompanied by the decrease in volume and increased by pressure); principle of microscopic ordering (increasing pressure increases the order of molecules at constant temperature); and isostatic principle (foods are subjected to uniform pressure from all directions and return to their original shape after release of pressure) [35].

High hydrostatic pressure processing has been used in the industries for the processing and preservation of meats and meat products. Its application can inactivate

Country	Irradiated meat and fish (tons)
The United States	8000
Belgium	5530
France	2789
The Netherlands	944
Indonesia	1008
Vietnam	14,200
Total	32,471

Table 1.
Quantity of meat and fish products that were irradiated in some countries.

pathogenic microorganisms and enzymes, increase shelf life, and maintain sensory quality [8, 35]. When compared to thermal treatments, the main advantage of this process is the maintenance of the sensorial and nutritional characteristics of the treated foods. With the wide application of this process, it is possible to develop value-added foods with better quality and shelf life compared to those produced in a conventional way [35].

Some of the important achievements in the treatment of meat product as chicken, pork, and beef processed by using HHP were presented in the works of [36–38], respectively.

The work of [36] applied HHP with control parameters 300 MPa for 5 min, at 20°C on fresh chicken breast fillets, and indicated that modified atmosphere packaging maintained their sensory attributes, color, tenderness, and microbiological quality. High hydrostatic pressure processing applied by Grossi et al. [37] indicated that treatment with 600 MPa for 6 min in pork affected the myofibrillar protein degradation pattern due to the increase of cathepsin activity. It was reported by Sanchez-Basurto et al. [38] that the HHP preserved raw meat over a longer-time period without significant difference of texture, tenderness, and color using the control parameters 172–620 MPa and 1–5 min in treatment.

In general, nonthermal technologies are not stand-alone techniques, and in order to improve the inactivation rates, several authors have proposed that HHP treatments are applied in combination with natural bioactive compounds, of which many originate from distinct natural sources [17].

In the work of Kalchayanand et al. [39], HHP treatment was used in roast beef samples inoculated with a mixture of clostridial spores that could be stored for 42 days at 4°C. It was observed that combined treatment of HHP and vacteriocin controlled the growth clostridium spores and extended the shelf life of roast beef for 84 days and can be stored at the same condition.

Different conditions for cooked ham were considered in the works of [40–42]. These studies compare samples submitted only to HHP treatment and in combination with some antimicrobials. The results reported by the authors generally show control of pathogen growth and increased shelf life.

A wide range of studies have been conducted to determine and enhance the efficacy in the combination of antimicrobial compounds with HHP treatments in the inactivation induced by pressure of pathogenic microorganisms. More details of the main recent results can be found in the work of [17].

2.3 Natural antioxidants

Lipid oxidation is the main cause for loss of sensory quality in meat products [28]. It is a chemical process that generates unpleasant odors, deterioration of the color, texture, and nutritive value of meat and meat products, which diminishes consumer acceptance, since the main attribute for evaluation of the food by the consumer is their appearance [1].

Meat proteins are also susceptible to oxidative reactions during heating and storage, and these reactions damage membranes and cellular functions, altering water retention, color, and reducing essential amino acids [1, 2].

To avoid the development of oxidative reactions, the industries use synthetic and natural antioxidants. Antioxidants have the functions of delaying or preventing oxidation processes, for example, in the elimination of free radicals. In this way, it will increase the shelf life and maintenance of food quality and safety [26, 43].

There are laws that regulate the use of antioxidants in food products. Sodium isocyanurate, butylhydroxyanisole (BHA), and butylhydroxytoluene (BHT) have

Source of antioxidant	Products	Main results	References
Plum	Irradiated turkey breast	Reduced lipid oxidation	[45]
	Precooked roast beef		[46]
Grape seed extract	Precooked sausages	Kept the odor of fresh meat cooked and longer flavor	[47]
Pomegranate	Chicken patties (hamburgers)	Reduced TBARS	[48]
	Cooled chicken meat		[49]
Rosemary	Frozen beef hamburgers	Red color stability during storage	[50]
	Irradiated frozen beef hamburgers	Reduced lipid oxidation	[25]
Oregano	Irradiated frozen beef hamburgers	Decreased lipid oxidation, but not as efficient when compared to rosemary extract	[25]

Table 2.
Natural extract used in meat and meat products.

been used in meat products. There are studies that show that synthetic antioxidants have the potential to cause toxicological effects, so it may be desirable to replace conventional antioxidants with natural antioxidants [25, 27, 43, 44].

These natural antioxidants are extracted in the form of extracts from different sources such as fruits (grapes and pomegranate), vegetables (broccoli and potatoes), herbs, and spices (tea, rosemary, oregano, cinnamon, sage, thyme, mint, ginger, and clove) [27]. The antioxidant, antimicrobial, and antifungal properties of these spices and extracts are mainly related to their bioactive components, such as phenolic compounds, flavonoids, vitamins, minerals, carotenoids, and phytoestrogens [1, 25].

Table 2 lists some studies that used natural extracts in meat and meat products.

In meat and poultry products, rosemary extract (*Rosmarinus officinalis*) is one of the most studied natural antioxidants, and its efficiency in turkey meat, ground beef, and pork has been reported [43].

The antioxidant activity of rosemary extract has been associated with the presence of several phenolics, such as carnosic acid, carnosol, rosmanol, and rosmaridiphenol, which has the function of breaking free-radical chains by electron and metal ion donation [25]. However, rosemary extract can be extracted from leaves and branches [27].

The use of synthetic and natural antioxidants helps to preserve the desirable characteristics of food. It is important to emphasize that when using a natural antioxidant, it is important to evaluate its impact on the sensorial analysis and quality of the final product [2, 43].

3. Conclusions

Consumers want to purchase quality meat products that are safe, nutritious, and natural, with appropriate appearance and flavor. To ensure the innocuity of products to consumers, various technologies and treatments can be used to achieve this result, such as nonthermal treatments, which ensure a safe and better quality product. In addition to these, treatments can use natural antioxidants to ensure a food with its natural characteristics. However, further studies are necessary to check the advantages and disadvantages of the beef hamburger irradiation process as well as in the use of combined processes may be involving the freezing, addition of natural antioxidants and irradiation.

Conflict of interest

The authors declare no conflict of interest.

Acronyms and abbreviations

BHA	butylhydroxyanisole
BHT	butylhydroxytoluene
FDA	Food and Drug Administration
HHP	high hydrostatic pressure
kGy	kiloGrays
TBARS	thiobarbituric acid reactive substances

Author details

Michelle Guimarães Horta, Fabiana Regina Lima, Carlos Alberto Gois Suzart and Poliana Mendes de Souza*
Universidade Federal dos Vales do Jequitinhonha e Mucuri, Brazil

*Address all correspondence to: poliana.souza@ict.ufvjm.edu.br

References

[1] Hygreeva D, Pandey M, Radhakrishna K. Potential applications of plant based derivatives as fat replacers, antioxidants and antimicrobials in fresh and processed meat products. Meat Science. 2014;**98**(1):47-57

[2] Ribeiro JS, Santos MJMC, Silva LKR, Pereira LCL, Santos IA, da Silva Lannes SC, et al. Natural antioxidants used in meat products: A brief review. Meat Science. 2018;**148**:181-188

[3] Cuppari L. Guia de nutrição: Nutrição clínica do adulto. In: Guia de nutrição: Nutrição clínica do adulto. 2005

[4] BRASIL. Ministério da Agricultura e do Abastecimento. Decreto n° 9013, de 29 de março de 2017. Decreto dispõe sobre o regulamento da inspeção industrial e sanitária de produtos de origem animal. Diário Oficial da União, Brasília. 2017

[5] Aymerich T, Picouet PA, Monfort JM. Decontamination technologies for meat products. Meat Science. 2008;**78**(1-2):114-129

[6] Stellato G, La Storia A, De Filippis F, Borriello G, Villani F, Ercolini D. Overlap of spoilage-associated microbiota between meat and the meat processing environment in small-scale and large-scale retail distributions. Applied and Environmental Microbiology. 2016;**82**(13):4045-4054

[7] Chai C, Lee SY, Oh SW. Shelf-life charts of beef according to level of bacterial contamination and storage temperature. LWT—Food Science and Technology. 2017;**81**:50-57

[8] Wang CY, Huang HW, Hsu CP, Yang BB. Recent advances in food processing using high hydrostatic pressure technology. Critical Reviews in Food Science and Nutrition. 2016;**56**(4):527-540

[9] Hugas M, Garriga M, Monfort J. New mild technologies in meat processing: High pressure as a model technology. Meat Science. 2002;**62**(3):359-371

[10] Nascimento MDGF, Oliveira CZF, Nascimento ER. Hambúrguer: Evolução comercial e padrões microbiológicos. Boletim do Centro de Pesquisa de Processamento de Alimentos. 2005;**23**(1):59-74

[11] Brasil. Ministério da Agricultura e do Abastecimento. Instrução Normativa n° 20, 31 jul. 2000. Dispõe sobre os regulamentos técnicos de identidade e qualidade de almôndega, de apresuntado, de fiambre, de hambúrguer, de kibe, de presunto cozido e de presunto, conforme consta dos anexos desta instrução normativa. Diário Oficial da União, Brasília. 2000

[12] Romans JR, Costello WJ, Jones K, Carlson CW, Ziegler PT, et al. The Meat We Eat. Danville: The Interstate Printers and Publishers Inc.; 2000

[13] Ojha KS, Tiwari BK, Kerry JP, Troy D. Beef. In: Caballero B, Finglas PM, Toldrá F, editors. Encyclopedia of Food and Health. Oxford: Academic Press; 2016. pp. 332-338. Available from: http://www.sciencedirect.com/science/article/pii/B9780123849472000568

[14] Pires RN, Caurio CF, Saldanha GZ, Martins AF, Pasqualotto AC. Clostridium difficile contamination in retail meat products in Brazil. The Brazilian Journal of Infectious Diseases. 2018;**22**(4):345-346

[15] Instituto Brasileiro de Geografia e Estatística (IBGE). Estatística da Produção Pecuária. Brasil, Rio de Janeiro: IBGE; 2018

[16] United States Department of Agriculture (USDA). Livestock and Poultry: World Markets and Trade. Foreign Agricultural Service, out. 2018. Available from: https://apps.fas.usda.gov/psdonline/circulars/livestock_poultry.pdf [Accessed: 27 May 2019]

[17] Oliveira TLC, Ramos AL, Ramos EM, Piccoli RH, Cristianini M. Natural antimicrobials as additional hurdles to preservation of foods by high pressure processing. Trends in Food Science & Technology. 2015;**45**(1):60-85

[18] Arvanitoyannis IS. Irradiation of Food Commodities: Techniques, Applications, Detection, Legislation, Safety and Consumer Opinion. Cambridge: Academic Press; 2010

[19] Farkas J. Irradiation for better foods. Trends in Food Science & Technology. 2006;**17**(4):148-152

[20] Diehl J. Food irradiation—Past, present and future. Radiation Physics and Chemistry. 2002;**63**(3-6):211-215

[21] Morehouse KM, Komolprasert V. Irradiation of food and packaging: An overview. In: ACS Symposium Series. Washington: ACS Publications; 2004. pp. 1-13

[22] Codex Alimentarius Commission (CAC). Codex general standard for irradiated foods. CODEX STAN. 2003:106-1983

[23] Brasil. Agência Nacional de Vigilância Sanitária. Resolução RDC n° 21, de 26 de janeiro de 2001. Dispõe sobre o regulamento técnico para irradiação de alimentos. Diário Oficial da União, Brasília. 2001

[24] Stefanova R, Toshkov S, Vasilev NV, Vassilev NG, Marekov IN. Effect of gamma-ray irradiation on the fatty acid profile of irradiated beef meat. Food Chemistry. 2011;**127**(2):461-466

[25] Trindade R, Mancini-Filho J, Villavicencio A. Natural antioxidants protecting irradiated beef burgers from lipid oxidation. LWT-Food Science and Technology. 2010;**43**(1):98-104

[26] Babuskin S, Babu PAS, Sasikala M, Sabina K, Archana G, Sivarajan M, et al. Antimicrobial and antioxidant effects of spice extracts on the shelf life extension of raw chicken meat. International Journal of Food Microbiology. 2014;**171**:32-40

[27] Shah MA, Bosco SJD, Mir SA. Plant extracts as natural antioxidants in meat and meat products. Meat Science. 2014;**98**(1):21-33

[28] Trindade R, Lima A, Andrade-Wartha E. Consumer's evaluation of the effects of gamma irradiation and natural antioxidants on general acceptance of frozen beef burger. Radiation Physics and Chemistry. 2009;**78**(4):293-300

[29] Codex Alimentarius Commission (CAC). General standard for the labelling of prepackaged foods. CODEX STAN. 1985:1-1985

[30] Kume T, Furuta M, Todoriki S, Uenoyama N, Kobayashi Y. Status of food irradiation in the world. Radiation Physics and Chemistry. 2009;**78**(3):222-226

[31] Chirinos RR, Vizeu DM, Destro MT, Franco BD, Landgraf M. Inactivation of Escherichia coli O157: H7 in hamburgers by gamma irradiation. Brazilian Journal of Microbiology. 2002;**33**(1):53-56

[32] Moura AFPd. Efeito da radiação ionizante sobre a oxidação do colesterol em hambúrgueres de frango e bovino congelados. In: Universidade de São Paulo. 2004

[33] Vieira VdS. Avaliação do potencial de aplicação do processo de irradiação na redução da população de Salmonella

sp. inoculada em hambúrguer de carne de frango: Aspectos sensoriais e vida de prateleira. In: Universidade de São Paulo. 2005

[34] Sun XD, Holley RA. High hydrostatic pressure effects on the texture of meat and meat products. Journal of Food Science. 2010;**75**(1):R17-R23

[35] Yordanov D, Angelova G. High pressure processing for foods preserving. Biotechnology & Biotechnological Equipment. 2010;**24**(3):1940-1945

[36] Rodríguez-Calleja J, Cruz-Romero M, O'sullivan M, García-López M, Kerry J. High-pressure-based hurdle strategy to extend the shelf-life of fresh chicken breast fillets. Food Control. 2012;**25**(2):516-524

[37] Grossi A, Gkarane V, Otte JA, Ertbjerg P, Orlien V. High pressure treatment of brine enhanced pork affects endopeptidase activity, protein solubility, and peptide formation. Food Chemistry. 2012;**134**(3):1556-1563

[38] Sanchez-BasurtoBE,Ramírez-GillyM, Tecante A, Severiano-Pérez P, Wacher C, Valdivia-Lopez MA. Effect of high hydrostatic pressure treatment on the preservation of beef meat. Industrial & Engineering Chemistry Research. 2011;**51**(17):5932-5938

[39] Kalchayanand N, Dunne CP, Sikes A, Ray B. Inactivation of bacterial endospores by combined action of hydrostatic pressure and bacteriocins in roast beef. Journal of Food Safety. 2003;**23**:219-233

[40] Marcos B, Jofre A, Aymerich T, Monfort JM, Garriga M. Combined effect of natural antimicrobials and high pressure processing to prevent Listeria monocytogenes growth after a cold chain break during storage of cooked ham. Food Control. 2008;**19**:76-81

[41] Vercammen A, Vanoirbeek KGA, Lurquin I, Steen I, Goemaere O, Szczepaniak S. Shelf-life extension of cooked ham model product by high hydrostatic pressure and natural preservatives. Innovative Food Science & Emerging Technologies. 2011;**12**:407-415

[42] Jofre A, Aymerich T, Garriga M. Assessment of the effectiveness of antimicrobial packaging combined with high pressure to control Salmonella sp. in cooked ham. Food Control. 2008;**19**:634-638

[43] Karre L, Lopez K, Getty KJ. Natural antioxidants in meat and poultry products. Meat Science. 2013;**94**(2):220-227

[44] Jayathilakan K, Sharma G, Radhakrishna K, Bawa A. Antioxidant potential of synthetic and natural antioxidants and its effect on warmedover-flavour in different species of meat. Food Chemistry. 2007;**105**(3):908-916

[45] Lee E, Ahn D. Quality characteristics of irradiated turkey breast rolls formulated with plum extract. Meat Science. 2005;**71**(2):300-305. DOI: 10.1016/j.meatsci.2005.03.017

[46] Gonzalez MN, Hafley B, Boleman R, Miller R, Rhee K, Keeton J. Antioxidant properties of plum concentrates and powder in precooked roast beef to reduce lipid oxidation. Meat Science. 2008;**80**(4):997-1004

[47] Kulkarni S, DeSantos F, Kattamuri S, Rossi S, Brewer M. Effect of grape seed extract on oxidative, color and sensory stability of a pre-cooked, frozen, re-heated beef sausage model system. Meat science. 2011;**88**(1):139-144

[48] Naveena BM, Sen AR, Vaithiyanathan S, Babji Y, Kondaiah N. Comparative efficacy of pomegranate

juice, pomegranate rind powder extract and BHT as antioxidants in cooked chicken patties. Meat Science. 2008;**80**(4):1304-1308 Available from: http://www.sciencedirect.com/science/article/pii/S0309174008002064

[49] Vaithiyanathan S, Naveena B, Muthukumar M, Girish P, Kondaiah N. Effect of dipping in pomegranate (Punica granatum) fruit juice phenolic solution on the shelf life of chicken meat under refrigerated storage (4 C). Meat science. 2011;**88**(3):409-414

[50] Milani LIG, Terra NN, Fries LLM, Kubota EH. Effect of the extract of persimmon (Diospyros kaki L.) cv.'Rama Forte'and rosemary oily extract (Rosmarinus officinalis L.) on the sensory characteristics and color stability of frozen beef burgers. Semina: Ciências Agrárias. 2012;**33**(3):1085-1094

Chapter 9

Importance and Applications of Ultrasonic Technology to Improve Food Quality

Maged E.A. Mohammed and Mohammed R. Alhajhoj

Abstract

Nutritional value and quality of food products are very important for a healthy life of human beings. Various modern thermal and nonthermal application technologies such as pulsed light, pulsed electric field, high and low hydrostatic pressure, microwave, and ohmic heating have been used to improve food products characteristics. In recent years, ultrasonic applications have been used for food processing. The ultrasonic is defined as sound waves with a frequency exceeding the human hearing limit. Based on the frequency range of ultrasonic waves, it can be used in many industrial applications including the processing of food. Applications of high-power ultrasonic with low frequency aim to improve the quality of food products. Low-power ultrasonic with high-frequency applications are used for nondestructive quality evaluation of physicochemical properties of food. The most important advantages of ultrasonic technologies are the low cost of food processing, low power consumption, simplicity compared to other technologies, suitability for the treatment of solid and liquid food, and environmental safeness and friendliness, thus becoming a promising technology for monitoring and improving quality of food products. The main objective of this chapter is to provide an overview of the principal and recent applications of ultrasonic waves to improve food product quality.

Keywords: ultrasonic applications, ultrasonic frequency, high-intensity, low-intensity, cavitation, ultrasonic equipment, food quality

1. Introduction

Improving the nutritional values and stability of quality is a very important parameter in food product quality assurance for a healthy life of human beings. Consumers are looking for fresh and good characteristics in their food with nutrient content and high sensorial quality. Now, consumers are more aware of the processing techniques used in the processing of their food, and they prefer natural products free of additives and chemicals. Therefore, there is a need for alternative technologies for food processing. Recently, various modern thermal and nonthermal technologies such as pulsed light, pulsed electric field, high and low hydrostatic pressure, microwave, ohmic heating, freezing, pasteurizing, ionizing radiation, etc. have been used to improve the physicochemical characteristics, extend the shelf life of food products, and control food quality by inactivating microorganisms at sublethal or ambient temperatures. One of the nonthermal technologies that can be used also is the application of ultrasonic (high-power and low-power ultrasonic with low and high frequency); especially it has shown a negligible effect on the nutrient value of food products [1, 2]. Applications of ultrasonic technology for food processing aim to offer consumers high-quality foods. The ultrasonic is considered to be a promising and emerging technology that can be used in food processing technology and many industrial applications by regulating frequency [3]. According to sound wave ranges used, the ultrasonic can be divided into low-power high-frequency ultrasonic and high-power low-frequency ultrasonic [4]. Low-power ultrasonic with high frequency is used for nondestructive quality evaluation of physicochemical characteristics of fruit, vegetables, and food products during processing or storage. The high-power ultrasonic with low frequency is used to improve the physicochemical properties of food products and in food processing such as humidification, hydrothermal treatments, extraction, drying, freezing, and inactivation of microorganisms of food products [3]. The ultrasonic technology has been also used in the industry of food products to develop many reliable and effective processing applications of food. The most common applications of ultrasonic in the industry of food include extraction of intracellular and material cell destruction. Depending on the ultrasonic intensity, the ultrasonic is used for the deactivation or activation of enzymes, homogenization and mixing, dispersion, stabilization, crystallization and dissolution, emulsification, hydrogenation, preservation, ripening, meat tenderization, oxidation, as a solid-liquid extraction adjuvant to accelerate and to improve the extraction, and atomization and degassing of food processing [5]. The objectives of ultrasonic research are to analyze and study the phenomena of undesirable and desirable degradation resulting from the applications of ultrasonic wave treatments in foods. The processing using ultrasonic may impact the chemical composition texture of foods [6].

Generally, ultrasonic applications are environmentally friendly and offer an advantage in the selectivity, yield, and productivity, with enhanced quality, reduced physical and chemical hazards, and short processing time. Before the commercialization of some food products such as vegetables and fruit, oils and fat, cocoa-sugar and coffee, meal and flours, dairy, and meat which are complex mixtures of proteins, sugars, lipids, vitamins, aromas, fibers, antioxidants, pigments, and mineral and organic compounds have to be processed and preserved using ultrasonic applications for food meals and to extraction of food ingredients [7]. The main purpose of this chapter is to provide an overview of the basic principles and current applications of low-intensity and high-intensity ultrasonic waves as a modern nonthermal technique for food product processing technology to improve its quality.

2. Overview of sound waves

The sound wave type is determined by its frequency. **Figure 1** shows the sound spectrum which displays the various frequencies present in a sound. "Infrasound" indicates a sound wave below the human hearing range. This frequency of sound is used by submarine sonar devices and whales. The frequency of the sound for the human hearing ranges from 20Hz to 20 kHz [8, 9]. The sound signal arises from many sources, e.g., the air turbulence or gases, passage through fluids, and by the impact of solid against another solid similar or non-similar. Because the sound is a natural phenomenon of waves, it may contain only one frequency as a sine wave with pure steady state (**Figure 2**) or contain complex frequencies such as the noise generated by many sound sources, e.g., machines and engines. The frequency of sound (f) is sound pressure times number. The sound frequency also may be identified by the frequency of angular (ω) expressed in radians per second as shown in Eq. (1). The period (T) is the time amount for a cycle of the single [10]:

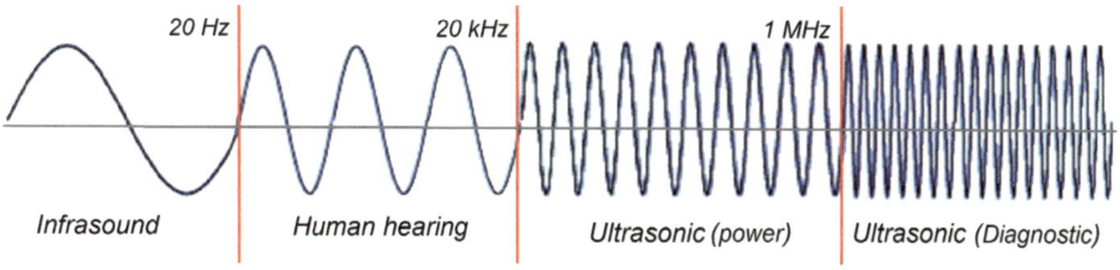

Figure 1.
Sound spectrum.

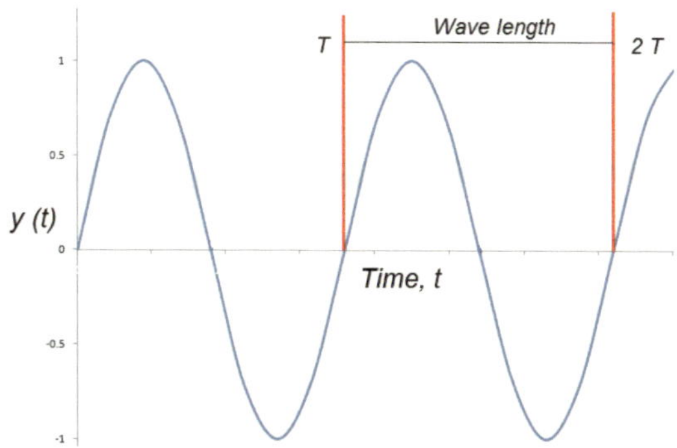

Figure 2.
The pure steady-state sine wave pulses.

$$\omega = 2\pi f = \frac{2\pi}{T} \tag{1}$$

Actually, the amplitude of the sound wave is strongly affected by the particles near the source of the sound waves, and on the contrary, the deeper particles are in the treated medium, the lower the sound wave amplitude. This reduction in sound wave amplitude at the deep is due to the attenuation produced by the treated medium. As a result, the sound amplitude versus wavelength distance is actually an exponentially sinusoid degenerate (**Figure 3**). The wavelength ((λ) is the distance between peaks of successive amplitude) is related to frequency (f) through the traveling wave velocity (c) as shown in Eq. (2) [11]:

$$\lambda = \frac{c}{f} \tag{2}$$

Ultrasonic is a wave of sound with a frequency greater than the human hearing limit. Ultrasonic is considered an energy form generated by a longitudinal mechanical

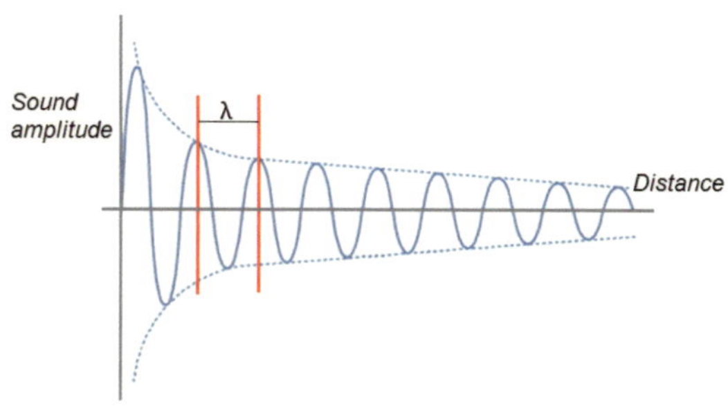

Figure 3.
Sinusoidal ultrasound wave.

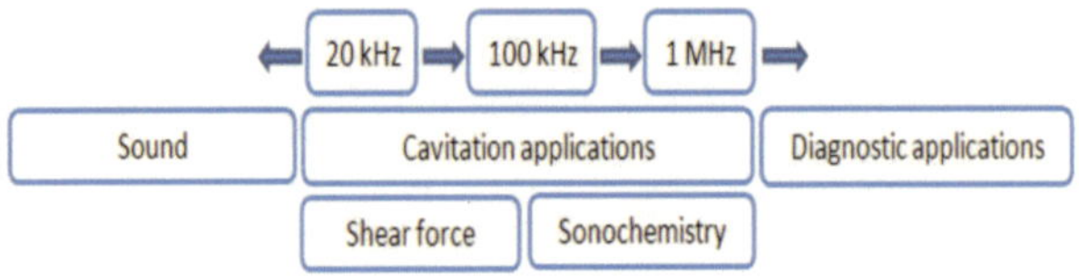

Figure 4.
Ultrasonic frequencies classification.

wave with one-dimensional propagation and frequency of vibration above 20,000 cycles per second (20 kHz) as shown in **Figure 4**. Ultrasonic waves can be categorized according to its frequency into two categories that are: (1) Low-frequency category which has frequency ranging from 20 to 1000 kHz. The applications of this category are used at high-power intensities in industrial applications, ultrasonic therapy, sonochemistry, and nanotechnology. (2) High-frequency category which has a frequency above 1 MHz and is being used at low-power intensities for nondestructive quality evaluation, imaging, and diagnostic applications [6, 12].

Use of ultrasonic application provides a good way to reach higher rates for the chemical and physicochemical process, shorter processing times and pathways of reaction. Interaction mechanisms between the product material and ultrasonic waves vary as a function of the input power of the ultrasonic. The pulse of ultrasonic speed depends on the acoustic properties of the medium of treated material. The speed of sound propagation in solid materials is higher than the sound propagation speed in liquids and greater in liquids than in gases [9].

3. Ultrasonic equipment

The main equipment of ultrasonic consists of a transducer, electrical power generator, and sound emitter devices. The emitter's function is to physically send the waves of ultrasonic to the medium. There are two types of ultrasonic systems used in the industry of food products: one using the bath as a traditional method and other using the horn as the sound emitter. The horn-based system is utilized in many applications from ultrasonic application in food processing and cleaning of plant surfaces for the process of food to application of ultrasonic for welding of metals [11].

3.1 Ultrasonic transducers

The transducer is the most important part of ultrasonic systems; the role of the transducer in the system is to generate the actual ultrasonic waves by converting the mechanical or electrical energy into sound energy at ultrasonic frequencies by vibrating mechanically. The ultrasonic transducer contacts to an electrical generator with 20 kHz frequency to transform electrical energy into ultrasonic energy by mechanical vibration at the same frequency (20 k cycles per second) [13]. The most applicable methods of ultrasound generation are carried out using ultrasonic transducers depending on the principle of the electrostrictive transformer. The principle of the methods is based on ferroelectric materials' elastic deformation within a high-frequency electrical field which results in molecules' polarized mutual attraction in the field. Then, the high-frequency alternating current is transmitted via two electrodes to ferroelectric material. After generating mechanical oscillation, the waves of sound are transmitted to the amplifier to generate the ultrasound [14].

The ultrasonic transducer is an electronic device that generates and receives the waves of sound. The transducer basically functions as a converter of energy, where it converts a form of acoustical energy into other energy forms (e.g., mechanical, electrical, or thermal energy). In addition, the transducer is reversible in either direction to convert electrical or mechanical energy to sound energy or vice versa. The most high-intensity ultrasonic generators are essentially magnetostrictive devices crystal oscillators in use. The categories of ultrasonic transducers fall into the following [10]:

1. Crystal oscillators are work through the effect of piezoelectric (reversible).

2. Magnetostrictive equipment are works based on the phenomenon of magnetostriction (reversible also).

3. Mechanical transducers that operate as generators and receivers.

4. Electromagnetic transducers are work based on the principle of the audio loudspeaker (but only work in the lower frequencies range of ultrasonic).

5. Other different types are thermal, optical, and chemical transducers.

6. Ultrahigh transducers that operate in the frequency range at megahertz or gigahertz.

Generally, the main transducers used in the most ultrasonic application can be summarized into three types: piezoelectric, magnetostrictive, and liquid-driven. The piezoelectric and magnetostrictive transducers convert magnetic and electrical energy into ultrasonic energy. The liquid-driven transducers depend on mechanical energy to generate ultrasonic energy [15].

3.2 Electrical generator

The electrical generators are used to supply the ultrasonic systems with the required electrical energy to drive the transducer. Generally, the electrical generator produces a suitable power rating for the ultrasound system and allows the power to be set only indirectly through current (I) and voltage (V) settings. The current represents the electron charge traversing an area over some time interval and measured in amp, the voltage represents the stored energy in the electrons and measured in volt, and the electrical power is the output of current and voltage. Electrical generators that are designed and operate in the low frequency ranged from 10 to 40 kHz

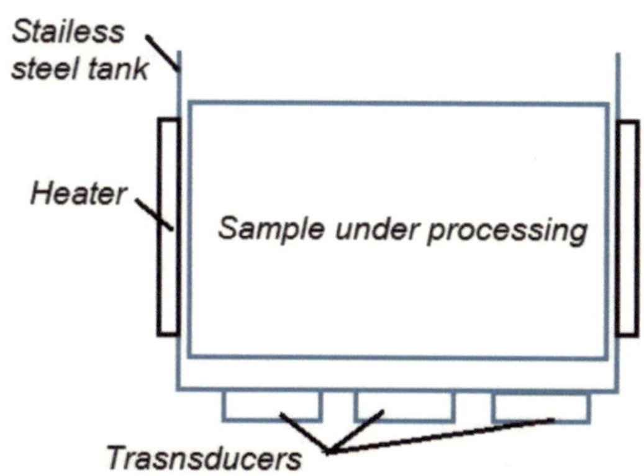

Figure 5.
Ultrasonic bath.

for ultrasonic generally focusing on industrial therapeutic applications, welding, cleaning, and disinfecting applications [11, 13].

3.3 Emitter (reactor)

The function of the emitter (reactor) is to radiate the waves of ultrasonic which are produced by the transducer into the treated medium. In addition, the role of the emitter may also be to amplify the ultrasonic vibrations when radiating them in some ultrasonic system. The main types of emitters are horns and baths; the horns often require a sonotrode to attach with the horn tip. The baths (**Figure 5**) usually consist of a stainless steel tank fixed with its base one or more transducers. The stainless steel tank holds a liquid case sample, and the transducers radiate ultrasonic directly into the sample [15].

4. Applications of power ultrasonic in food processing

Although the ultrasonic has been used in the twentieth century, most of the new and improved ultrasonic applications has reached practically only in the past few years. Ultrasonic applications can be classified into two categories as high intensity and low intensity. High-intensity applications deliberately affect the contents of the propagation medium. Uses of high intensities include liquid atomization, material machining, medical surgery and therapy, material cleaning, plastics and metals welding, biological cell disruption, and material homogenization. Low-intensity applications carry the objective of transmitting energy through a medium in order to convey information through the treated medium or to obtain information about the medium. Uses of low intensities include nondestructive testing, medical diagnosis, elastic property measurements of materials and agricultural products, and acoustical holography. Nowadays, ultrasonic application technology has extremely affected the meat industry, with a controlling role in the classification of the product quality. It is being used to measure the fat layer thickness in live animals, and it is also utilized to predict carcass traits as a livestock management part, and it has been used to improve homogenized milk quality. In addition, the ultrasonic application technology is utilized in the pest control that includes the expulsion or killing of insects [10, 11]. The potential uses of ultrasonic applications technology for improving the nutritional and quality aspects of food have been highlighted by Ashokkumar [16]. The ultrasonic application technology offers a huge potential to bioprocessing industries and foods. Developing custom-made and new equipment is an issue to be addressed by food technologists, physicists, and engineers [16].

In addition, ultrasonic applications have been used for food processing as an important alternative processing method of conventional thermal. Ultrasonication process can preserve and pasteurize food products by inactivation of microorganisms and many enzymes at normal conditions of temperature to guarantee the safety and stability of foods for improving food quality. The changes in ultrasonic physical properties, such as attenuation and scattering caused by treated food product materials, have been also used in applications of food quality assurance [17]. The potential applications of ultrasonic are not only affected by the medium (gas, liquid, solid, or supercritical) but also the treatments variables (flow regime, temperature, ultrasonic intensity, etc.) and the structure of product which could affect the magnitude of the changes induced by ultrasonic processing [18]. Ultrasound can be divided into different frequency ranges. Most ultrasonic applications in the food processing technology involved nondestructive measurements which referred especially to the assessment of product quality; such applications use low power less than 1 W/cm^2

and high-frequency ultrasonic of 100 kHz to 1 MHz. Low-intensity ultrasonic is commonly applied as an analytical method to provide information on the food product's physicochemical properties such as acidity, ripeness, firmness, content of sugar, etc. The high power levels used (typically in the range 10–1000 W/cm^2) with low frequency (16–100 kHz) are used to make physical or chemical changes in the food to improve its properties [11, 19].

Generally, the ultrasonic applications are separated into two categories: the first category is low-intensity ultrasonic (called nondestructive or high-frequency ultrasound), and the second category is high-intensity ultrasonic (called low-frequency or power ultrasound) [20].

4.1 Applications of low-intensity ultrasonic

Low-intensity high-frequency ultrasonic is a nondestructive technique which is applied for detection purposes and provides information about the physicochemical characteristics of food products such as structure, firmness, composition, flow rate, physical state, etc. [21]. The action of ultrasonic waves is dependent on the input power. So the low-power ultrasonic is considered a noninvasive nondestructive method, and it is a useful technique for characterizing the physicochemical properties of food products, determining the food components type and contents, and measuring the emulsions droplet size. Irradiation of food products by low-power ultrasonic did not create any physical changes, at variance the high-power ultrasonic created the changes [9]. It is also used as a processing method in the industry of food to describe the components of food products, often in line with quality assurance. The nondestructive test basically is done by sending waves of ultrasonic through the medium without causing any permanent electrical, chemical, or physical changes in the food products. This is due to the use of too low ultrasonic intensity (<1 W/cm^2), so there is no change in the foods by using this [11, 20, 21].

When ultrasonic waves pass through the medium, the particles in the medium oscillate mechanically in response to the low-intensity (low-energy) ultrasound. After that, the particles exposed to the waves of ultrasonic simply return to their position of equilibrium when the ultrasonic source is stopped. The distance to the location of reflection can be calculated by measuring the attenuation coefficient and frequency properties of ultrasonic to evaluate the physicochemical properties and to allow detection of compositional changes in the food products [11, 21]. In using low-intensity ultrasonic to characterize vegetable and fruit properties, there must be a relationship between the property to be measured and any measurable parameter of ultrasonic (e.g., impedance, attenuation, or velocity). The particular parameter that often influences the properties of ultrasonic in vegetables and fruits is the presence of intercellular air spaces that causes a resonant phenomenon over ultrasonic frequencies in a wide range. The appropriate frequency which transmits normally through vegetables and fruits is above 1 MHz at low intensity to avoid the damage in plant tissue [22].

On the other hand, there are other indirect applications for high-frequency ultrasonic in food processing area such as applications of ultrasonic in humidifiers or misting devices which are used in humidification or hydration of fresh fruit and vegetables or humidification systems of meat in the cold storage rooms for improving the quality of the product and decreasing the weight loss during the storage period. The operating principle of ultrasonic humidifiers depends on converting the electrical energy into periodically mechanical vibration by piezoelectric transducers and horn, which vibrates at high frequencies. The piezoelectric transducers are placed at the bottom of the water in order to produce high-frequency waves that propagate upward into the water. Then the ultrasonic wave rarefaction cycle causes cavitation; in addition, the water over the piezoelectric transducer will produce a wavy layer. If the ultrasonic waves have enough energy that can overcome the water surface tension, then droplets will be generated from the water top surface. When the vibrating surface amplitude is increased to a level that the ultrasonic waves collapse and are unstable, the droplets will be ejected away from the water surface into a mist. The droplet's size is dependent on the frequency of vibration and water depth above the piezoelectric transducers [23, 24].

The diameter of the atomized droplets is calculated based on the properties of the ultrasonic generator by Eq. (3) [25–27]:

$$D_d = \alpha_g \sqrt[3]{8\pi \times \sigma_s/\rho_s \times f^2} \tag{3}$$

where D_d represents the droplet diameter, f is ultrasonic frequency, ρ is a liquid density, σ is a surface tension, and α_g is a dimensionless constant ($\alpha_g \approx 0.4$).

Generally, low-intensity ultrasonic applications can invaluably improve quality control in food production and monitor the changes that occur during humidification, emulsifying, freezing, or drying of food products. Some food manufacturers use nondestructive ultrasonic applications to locate foreign particles such as organic residues, bacterial infections, or glass in solid and liquid food products during and even after food packaging [28]. Low-intensity ultrasonic has been used successfully at ultrasonic wave frequency of 150 kHz as a noninvasive and nondestructive means of evaluating the commercial poultry egg quality at different conditions of storage using the velocity ultrasound phase within the material of eggs to recognize the differences between the aged and fresh eggs [29].

Low-intensity ultrasonic applications are considered one of the efficient tools for nondestructive quality evaluation of fresh fruits and vegetables. These applications are characterized as a reliable and fast technique for correlating fruit and vegetable properties and specific indices of quality with the different growth stages, after maturation, during storage, and after storage to be ready for marketing and consumption while ensuring its quality. Commercial application of ultrasonic applications will be useful to consumers and growers due to the public demand for high-quality and uniform agricultural products [30]. High-frequency ultrasonic technique using a contact transducer of 100 kHz as a nondestructive tool to determine fruit quality of navel oranges was applied successfully after fruit harvesting with a high accuracy level. Water content and density of the fruit can be determined accurately regardless of the other physical properties such as maturity, size, and the peel uniformity by isolating the results section which relates straight to the fruit acoustic properties. There is a high level of correlation between orange firmness and the reflected energy quantity of ultrasonic. Using ultrasonic technique, substandard individual fruit can be identified and sorted to be discarded at any harvest time and during processing or in a storage room. On the contrary, the methods of traditional destructive can be applied only on a limited sample of fruit after harvesting [31]. The measurements of the ultrasonic velocity (high-frequency) and attenuation which was conducted at 25 MHz on samples of mango juice showed a big variability with a maturity of fruit at picking and after picking at ripening stage in relation to texture of fruit, the content of total soluble solids (TSS), and changes in biochemical composition [32]. Many research has been done on nondestructive applications of ultrasonic technologies in food processing, but further future research is needed in this area in order to develop new automated ultrasonic equipment.

4.2 Applications of high-intensity ultrasonic

Applications of high-intensity ultrasonic or power ultrasonic are used to change the physical or chemical properties of food products as well as to promote the reactions of chemicals, produce emulsions, inhibit enzymes, disrupt cells, crystallization processes modification, etc. [21]. The use of high-intensity low-frequency ultrasonic waves generating sonotrodes was initially proposed for cleaning, emulsification, and bacterial lysing. The high-intensity ultrasonic wave (high-power) equipment using sonotrodes operating was further developed for processes of chemicals (up to 6 kW). In recent, ultrasonic systems are developed to generate high mega-sonic frequencies of ultrasound (400 kHz) with a high power level (>100 W). Therefore, the high-intensity ultrasonic wave is suitable for many applications in food products [33].

The high-intensity ultrasonic fundamental effect on the fluid material is for effective hydrostatic pressure on the medium and the imposition of acoustic pressure. The acoustic pressure (P_a) is a sine wave dependent on the ultrasonic frequency (f), time (t), and the wave pressure amplitude at the maximum ($P_{a\text{-}max}$) [Eq. (4)]. The maximum wave pressure is proportional to the transducer power input [34]:

$$P_a = P_{a\,max} \sin(2\pi f t) \tag{4}$$

The application of high-intensity ultrasonic (high power level = 75 W) was developed and tested to assist in convective heat transfer during food drying. The application of ultrasonic is based on the ultrasonic energy transmission through airborne contacts and solid contact series between the ultrasonic transducer and the tray of the food product as a vibration surface of ultrasonic transmitting. The slices of apple were dried using this method without compromising the quality of the product. The results indicated that using the ultrasonic application during apple drying led to the following: processing time was accelerated, consumption energy was reduced, production throughput was increased, and the quality of the product was not affected by ultrasonic processing. The results also indicated that the ultrasonic treatments led to improve the convective drying process efficiency when using high-power ultrasonic at low temperature. These results are very useful at the need to dehydrate heat-sensitive products effectively or to decrease food drying time in order to preserve the physicochemical and nutritional properties of food products [35]. Pasteurization of many food products by an ultrasonic application at 50°C has a preserving potential on the food quality in terms of color, flavor, and physicochemical properties compared to the techniques of conventional pasteurization at high temperatures [36]. The propagation of ultrasonic in a medium causes chemical and physical impacts, and these impacts have been used to improve the efficiency of the operations of various food processing technologies, and it has been also used as diagnostic technology in food quality control. The high-intensity ultrasonic application was applied to control ice crystal's size distribution in low-temperature processes and related applications such as thawing, freezing, freeze-drying, and freeze concentration. It has been led to improve the freezing process efficiency, accelerate the freezing rate, and ensure frozen food quality [37].

High-intensity ultrasonic is being applied as an efficient preservation tool in fields of food processing for fruits and vegetables, honey, cereal products, proteins, gels, enzymes, cereal technology, dairy technology, water treatment, microbial inactivation, etc. [38]. In a previous study, the researchers have studied the effects of high-intensity ultrasonic at different levels of power ultrasound of 0, 200, 400, and 600 W as nonthermal processing on microbial inactivation (aerobic mesophilic, molds, yeasts, and coliforms), microstructure (particle size distribution and optical microscopy), rheology, color, and kinetic stability of the inulin-enriched whey beverage. The result obtained by applying ultrasonic power of 600 W was comparable to applying a high temperature of 75°C at short treatment time of 15 s concerning the total microbial inactivation. In addition, the high-intensity ultrasonic was better than the high-temperature short-time ultrasonic in improving kinetic stability of beverage, decreasing consistency and viscosity, avoiding phase separation, disrupting fruit and milk cells, and decreasing particle size. Therefore, nonthermal processing by high-intensity ultrasonic seems to be a promising technology for the production of probiotic dairy beverages. However, further future studies concerning the ultrasonic application effect on nutritional properties of this product must be evaluated before marketing [39]. Sterilization and improved emulsification can be conducted at lower temperatures than conventional treatments at high temperatures using high-intensity ultrasonic to produce a stable food product by retaining the useful bioactive ingredients and preventing spoilage of treated food. Applications of high-intensity ultrasonic in the fractionation of fat, dairy beverages production, and disruption of casein offer the potential of decreased treatment times; properties of the possible product have more advantages than those produced through conventional thermal techniques. Therefore, using ultrasonic applications in this area will lead to economic savings to producers in terms of producing value-added products and processing times and temperature. The consumers were satisfied with ultrasonic application studies for processing of food products to improve the quality of final products in terms of flavor, color, texture, and other physicochemical characteristics [40].

High-intensity ultrasonic treatment is a good process to inactivate enzymes and microorganisms at combined pressure and heat treatments as a hurdle technology. This combination is a successful application in lower temperatures for the inactivation process which provides a good solution for food product producers to secure fresh-like foods [41]. The impact of power ultrasound on the fruit and vegetable quality during drying and pre-treatment has been assessed. The indicators of fruit quality such as the losses of leaching, rehydration capacity, shrinkage of fruit, and the final product's organoleptic characteristics have been also evaluated. The result showed that enzyme inactivation and leaching losses during blanching using high-intensity high-power ultrasonic at low temperature are similar to the result found using conventional treatments, but there is a significant reduction in the ultrasonic treatment time. Ultrasonic application in the drying of strawberries and carrots produces a highly significant reduction in the time of processing while providing high-quality final products. The final products' quality was equivalent or superior to final products obtained in convective dryer prototype under similar conditions, was higher than marketed products, and was similar to the produced products by freeze-drying [42]. The impact of low-frequency high-power ultrasound (40 kHz, 130 W) on bean in terms of kinetics of hydration and cooking times was studied. Treatment of bean samples by ultrasonic waves for 30 min at 30°C 30 min occurs a significant increase in the effective diffusivity up to 45 times and reduces the time which obtains the equilibrium moisture content by 58.8% and the reduction percentage in cooking time reached 43% [43].

Generally, high-power ultrasonic has become an efficient technique for some commercial applications, such as homogenization, emulsification, crystallization, extraction, dewatering, low-temperature pasteurization, deforming, degassing, viscosity alteration, reduction of particle size, and inactivation or activation of enzymes. In addition, due to the need for inactivation of enzymes and microorganisms without destroying food nutrients, the high-power ultrasound applications are the best processing methods as a nonthermal alternative method to thermal processing treatments for food product preservation. This is due to continuous development and improvement in the design and manufacturing of ultrasonic equipment, but high-power ultrasonic for food processing like most innovative technologies in this field is not an effective technique for large-scale commercial application. Therefore, there is a need to conduct research on high-power ultrasonic for it to become an efficient large-scale commercial technology for processing food products [36, 44].

5. Applications of cavitation

The cavitation phenomenon (liquid rupture) is easily observed in water boiling, turbines, hydrofoils, and in seawater in the proximity of a rotating propeller of the ship. It happens in those liquid regions that are subject to rapidly vacillating pressures with high amplitude. Cavitations also happen in a liquid exposed to high-energy ultrasonic, considering that the sound travels through a small volume of fluid or water. During the negative half of the pressure cycle, the liquid is exposed to tensile stress, and during the positive half of the pressure cycle, the liquid is exposed to compression stress. Therefore, the bubbles entrapped in the liquid will extend and retract alternatively. When the amplitude of pressure is sufficiently large and the bubble initial radius is minimal than the critical value, R0 is given using the following equation [10]:

$$R_o = 1/\omega\sqrt{3\gamma\left(\left(P_o + \frac{2T_{st}}{R_o}\right)/\rho\right)} \tag{5}$$

where ω is a signal angular frequency, P_o is the hydrostatic pressure in a liquid, γ is a principal specific heat ratio of the gas in a bubble, T_{ts} is a surface tension at the bubble surface, and ρ is the intensity of liquid.

The sound pressure load (<10 Pa) exerted on the ear of human is very small, but the pressure of ultrasonic (MPa) in liquids can be high enough to create the cavitation phenomenon (can destroy the treated medium). Ultimately, the cavitations lead to free radical production and sonochemical that react chemically with media (liquid) and also lead to the destruction of microbiological cells [11]. The ultrasound passage in liquid products generates a physical effect and mechanical agitation due to acoustic cavitation [16]. The food industry has usually depended on the heating methods for enzyme and microorganism inactivation for preservation of food products. Despite thermal method actually leading to destroy some spores, kill microorganisms, and inactivate some enzymes, food may lose their organoleptic and nutritional properties during the process. On the contrary, the inactivation mechanism using the ultrasonic application depends on the generation of physical forces due to the phenomenon of cavitation [17]. Transmitted, dispersed, and reflected pulses of acoustic can be used in food product quality assurance. Using ultrasonic application for enzymatic inactivation of some food products is very important for the preservation of quality which is a requisite for secure food material stabilization. The physical and chemical forces generated by ultrasonic cavitation raise severe damage to the microorganism's cell wall, leading to microorganism inactivation. In addition, ultrasonic cavitation effects in liquid foods lead to disrupting the functional and structural components up to microorganisms cell lysis [17]. The applications of ultrasonic that are used for flaw detection in food processing for quality assurance of food products must be designed with ensuring that no cavitation possibly occurs. On the contrary, there are other applications of ultrasonic, depending on inertial cavitation to produce desirable changes in food products. These changes are produced by cavitation, such as microorganism inactivation and release of nutritional compounds and oils through the erosion of the cellular structure of the treated product cell [11]. The released energy during cavitation has a great ability to improve food products' safety by destroying the pathogenic and food spoilage microorganisms and foreign material detection in food products. Although the applications of cavitation are well applied in many different industries other than food processing, the application of cavitation in processing of dairy products and its ingredients is recently gaining much attention, and it has a large potential to become a promising method in the near future in dairy product processing area such as reduction of viscosity, homogenization, making of yogurt, cream, and cheese, waste management, microbial inactivation, food safety, etc. Power ultrasound cleaning application at low frequency generally operates between 20 and 50 kHz. The cleaning effect of ultrasound depends on cavitation. Increasing the cavitation in cleaning liquid increases the ultrasonic cleaning effect. The most important parameters affecting the cavitation are ultrasonic frequency and temperature [41, 45].

6. Conclusions

Ultrasonic applications can be considered promising and applicable as a green technology for food safety and quality assurance purposes of food products. Ultrasonic applications are divided into two categories according to its intensity and frequency: high frequency with low intensity (power ultrasound) and low frequency with high intensity (nondestructive). High-intensity applications deliberately affect the contents of the propagation medium. Uses of high intensities offer an advantage in the selectivity, yield, and productivity, with enhanced quality, reduced physical and chemical hazards, and short processing time. Before the commercialization of some food products such as vegetables and fruit, oils and fat, cocoa-sugar and coffee, meal and flours, dairy, and meat which are complex mixtures of proteins, sugars, lipids, vitamins, aromas, fibers, antioxidants, pigments, and mineral and organic compounds have to be processed and preserved using ultrasonic applications for food meals and to extraction of food ingredients. Low-intensity applications are used for nondestructive quality evaluation of food products. The physical and chemical forces generated by ultrasonic cavitation raise severe damage to the cell wall of microorganisms, leading to their inactivation. A major advantage of the ultrasonic applications in food processing is that it is perceived as benign by the consumers. On the contrary, other processing technologies such as gamma radiation, microwaves, ohmic heating, and pulsed electric field can be cautiously considered by some of the population. Generally, the sound waves are considered nontoxic, safe, and environmentally friendly; this gives the use of ultrasonic major advantage over other modern processing techniques. In addition, it is characterized by the low cost of construction, low power consumption, simplicity compared to other technologies, and suitability for solid and liquid food products. Despite conducting a lot of research on applications of ultrasonic technologies for food products, there is still a need for more future research in order to utilize this technology on a fuller industrial scale to produce high-quality and safe food products.

Acknowledgements

The authors gratefully acknowledge the financial support for the project number (DPRC-7-2018) by the Date Palm Research Center of Excellence, King Faisal University, KSA.

Conflict of interest

The authors declare no conflict of interest.

Author details

Maged E.A. Mohammed[1,2*] and Mohammed R. Alhajhoj[1]

1 Date Palm Research Center of Excellence, King Faisal University, KSA

2 Agricultural Engineering Department, Faculty of Agriculture, Menoufia University, Egypt

*Address all correspondence to: maged3929@yahoo.ca

References

[1] Mohamed ME, Eissa AH. Pulsed electric fields for food processing technology. In: Eissa AH, editor. Structure and Function of Food Engineering. Rijeka, Croatia: IntechOpen; 2012. pp. 275-306. DOI: 10.5772/48678.ch11

[2] Mohammed MEA, Eissa AHA, Aleid SM. Application of pulsed electric field for microorganisms inactivation in date palm fruits. Journal of Food and Nutrition Research. 2016;**4**:646-652. DOI: 10.12691/jfnr-4-10-3

[3] Awad TS, Moharram HA, Shaltout OE, Asker D, Youssef MM. Application of ultrasound in analysis, processing and quality control of food. A review. Food Research International. 2012;**48**:410-427. DOI: 10.1016/j.foodres.2012.05.004

[4] Dolatowski ZJ, Stadnik J, Stasiak D. Applications of ultrasound in food technology. Acta Scientiarum Polonorum. Technologia Alimentaria. 2007;**6**(3):88-99

[5] Gallo M, Ferrara L, Naviglio D. Application of ultrasound in food science and technology: A perspective. Food. 2018;7:164. DOI: 10.3390/foods7100164

[6] Alarcon-Rojo AD, Peña-González E, García-Galicia I, Carrillo-López L, Huerta-Jiménez M, Reyes-Villagrana R, et al. Ultrasound application to improve meat quality. In: Díaz AV, García-Gimeno RM, editors. Descriptive Food Science. Rijeka, Croatia: IntechOpen; 2018. pp. 153-172. DOI: 10.5772/intechopen.77973

[7] Chemat F, Khan MK. Applications of ultrasound in food technology: Processing, preservation and extraction. Ultrasonics Sonochemistry. 2011;**18**:813-835. DOI: 10.1016/j.ultsonch.2010.11.023

[8] Feng H, Barbosa-Cánovas GV, Weiss J. Ultrasound Technologies for Food and Bioprocessing. 1st ed. New York: Springer; 2011. DOI: 10.1007/978-1-4419-7472-3

[9] Kasaai MR. Input power-mechanism relationship for ultrasonic irradiation: Food and polymer applications. Natural Science. 2013;**5**:14-22. DOI: 10.4236/ns.2013.58A2003

[10] Raichel DR. The Science and Applications of Acoustics. New York: Springer Science & Business Media; 2006

[11] Bermúdez-Aguirre D, Mobbs T, Barbosa-Cánovas GV. Ultrasound applications in food processing. In: Feng H, Barbosa-Cánovas GV, Weiss J, editors. Ultrasound Technologies for Food and Bioprocessing. 1st ed. New York: Springer; 2011. pp. 65-105. DOI: 10.1007/978-1-4419-7472-3

[12] Tyagi VK, Lo SL, Appels L, Dewil R. Ultrasonic treatment of waste sludge: A review on mechanisms and applications. Critical Reviews in Environmental Science and Technology. 2014;**44**:1220-1288. DOI: 10.1080/10643389.2013.763587

[13] Mason TJ. Power ultrasound in food processing—The way forward. In: Povey MJW, Mason TJ, editors. Ultrasound in Food Processing. London: Blackie Academic & Professional; 1998. pp. 103-126

[14] Knorr D, Zenker M, Heinz V, Lee DU. Applications and potential of ultrasonics in food processing. Trends in Food Science and Technology. 2004;**15**:261-266. DOI: 10.1016/j.tifs.2003.12.001

[15] Povey MJW, Mason TJ. Ultrasound in Food Processing. London: Blackie Academic & Professional; 1998. p. 271

[16] Ashokkumar M. Applications of ultrasound in food and bioprocessing. Ultrasonics Sonochemistry. 2015;**25**:17-23. DOI: 10.1016/j.ultsonch.2014.08.012

[17] Chandrapala J, Oliver C, Kentish S, Ashokkumar M. Ultrasonics in food processing—Food quality assurance and food safety. Trends in Food Science and Technology. 2012;**26**:88-98. DOI: 10.1016/j.tifs.2012.01.010

[18] Cárcel JA, García-Pérez JV, Benedito J, Mulet A. Food process innovation through new technologies: Use of ultrasound. Journal of Food Engineering. 2012;**110**:200-207. DOI: 10.1016/j.jfoodeng.2011.05.038

[19] Soria AC, Villamiel M. Effect of ultrasound on the technological properties and bioactivity of food: A review. Trends in Food Science and Technology. 2010;**21**:323-331. DOI: 10.1016/j.tifs.2010.04.003

[20] Mason TJ. Sonochemistry and sonoprocessing: The link, the trends and (probably) the future. Ultrasonics Sonochemistry. 2003;**10**:175-179. DOI: 10.1016/S1350-4177(03)00086-5

[21] McClements JD. Advances in the application of ultrasound in food analysis and processing. Trends in Food Science and Technology. 1995;**6**:293-299. DOI: 10.1016/S0924-2244(00)89139-6

[22] McClements DJ, Gunasekaran S. Ultrasonic characterization of foods and drinks: Principles, methods, and applications. Critical Reviews in Food Science and Nutrition. 1997;**37**:1-46. DOI: 10.1080/10408399709527766

[23] Yeo LY, Friend JR, McIntosh MP, Meeusen ENT, Morton DAV. Ultrasonic nebulization platforms for pulmonary drug delivery. Expert Opinion on Drug Delivery. 2010;7:663-679. DOI: 10.1517/17425247.2010.485608

[24] Phanphanit P. Experimental and computational study of an ultrasonic atomizer [thesis]. Faculty of Engineering and Physical Sciences, University of Manchester; 2011

[25] Gagnon AM. Fabrication of an ultrasonic nebulizer: Rate of flow and performance studies. Studies in Conservation. 2019;**64**:1-12. DOI: 10.1080/00393630.2019.1605124

[26] Rajan R, Pandit A. Correlations to predict droplet size in ultrasonic atomization. Ultrasonics. 2001;**39**:235-255. DOI: 10.1016/S0041-624X(01)00054-3

[27] Rodes CTRG, Smith T, Crouse R, Ramachandran G. Measurements of the size distribution of aerosols produced by ultrasonic humidification. Aerosol Science and Technology. 1990;**13**:220-229. DOI: 10.1080/02786829008959440

[28] Hæggström E, Luukkala M. Ultrasound detection and identification of foreign bodies in food products. Food Control. 2001;**12**:37-45. DOI: 10.1016/S0956-7135(00)00007-4

[29] Aboonajmi M, Akram A, Nishizu T, Kondo N, Setarehdan SK, Rajabipour A. An ultrasound based technique for the determination of poultry egg quality. Research in Agricultural Engineering. 2010;**56**:26-32. DOI: 10.17221/18/2009-RAE

[30] Mizrach A. Ultrasonic technology for quality evaluation of fresh fruit and vegetables in pre-and postharvest processes. Postharvest Biology and Technology. 2008;**48**:315-330. DOI: 10.1016/j.postharvbio.2007.10.018

[31] Morrison DS, Abeyratne UR. Ultrasonic technique for non-destructive quality evaluation of oranges. Journal of Food Engineering. 2014;**141**:107-112. DOI: 10.1016/j.jfoodeng.2014.05.018

[32] Valente M, Prades A, Laux D. Potential use of physical measurements including ultrasound for a better mango fruit quality characterization. Journal of Food Engineering. 2013;**116**:57-64. DOI: 10.1016/j.jfoodeng.2012.11.022

[33] Madhu B, Sai Srinivas M, Srinivas G, Jain S. Ultrasonic technology and its applications in quality control, processing and preservation of food: A review. Current Journal of Applied Science and Technology. 2019;**32**:1-11. DOI: 10.9734/CJAST/2019/46909

[34] Muthukumaran S, Kentish SE, Stevens GW, Ashokkumar M. Application of ultrasound in membrane separation processes: A review. Reviews in Chemical Engineering. 2006;**22**:155-194. DOI: 10.1515/REVCE.2006.22.3.155

[35] Sabarez HT, Gallego-Juarez JA, Riera E. Ultrasonic-assisted convective drying of apple slices. Drying Technology. 2012;**30**:989-997. DOI: 10.1080/07373937.2012.677083

[36] Patist A, Bates D. Ultrasonic innovations in the food industry: From the laboratory to commercial production. Innovative Food Science & Emerging Technologies. 2008;**9**:147-154. DOI: 10.1016/j.ifset.2007.07.004

[37] Cheng X, Zhang M, Xu B, Adhikari B, Sun J. The principles of ultrasound and its application in freezing related processes of food materials: A review. Ultrasonics Sonochemistry. 2015;**27**:576-585. DOI: 10.1016/j.ultsonch.2015.04.015

[38] Majid I, Nayik GA, Nanda V. Ultrasonication and food technology: A review. Cogent Food & Agriculture. 2015;**1**:1071022. DOI: 10.1080/23311932.2015.1071022

[39] Guimarães JT, Silva EK, Alvarenga VO, Costa ALR, Cunha RL, Sant'Ana AS, et al. Physicochemical changes and microbial inactivation after high-intensity ultrasound processing of prebiotic whey beverage applying different ultrasonic power levels. Ultrasonics Sonochemistry. 2018;**44**:251-260. DOI: 10.1016/j.ultsonch.2018.02.012

[40] Paniwnyk L. Applications of ultrasound in processing of liquid foods: A review. Ultrasonics Sonochemistry. 2017;**38**:794-806. DOI: 10.1016/j.ultsonch.2016.12.025

[41] Ünver A. Applications of ultrasound in food processing. Green Chemical and Technological Letters. 2016;**2**:121-126. DOI: 10.18510/gctl.2016.231

[42] Villamiel M, Gamboa J, Soria AC, Riera E, García-Pérez JV, Montilla A. I mpact of power ultrasound on the quality of fruits and vegetables during dehydration. Physics Procedia. 2015;**70**:828-832. DOI: 10.1016/j.phpro.2015.08.169

[43] Ulloa JA, Enríquez López KV, Contreras Morales YB, Rosas Ulloa P, Ramírez Ramírez JC, Ulloa Rangel BE. Effect of ultrasound treatment on the hydration kinetics and cooking times of dry beans (phaseolus vulgaris). CyTA Journal of Food. 2015;**13**(4):588-596. DOI: 10.1080/19476337.2015.1024173

[44] Ercan S, Soysal Ç. Use of ultrasound in food preservation. Natural Science. 2013;**5**:5-13. DOI: 10.4236/ns.2013.58A2002

[45] Sutariya S, Sunkesula V, Kumar R, Shah K. Emerging applications of ultrasonication and cavitation in dairy industry: A review. Cogent Food & Agriculture. 2018;**4**:1549187. DOI: 10.1080/23311932.2018.1549187

Section 5

Factors That Influence Food Processing

Chapter 10

Effect of Polyethylene Glycol 3350 on the Handling Properties of Low Salt Wheat Dough Formulations

Andrea K. Stone, Aleksandar Yovchev, Pierre J. Hucl, Martin G. Scanlon and Michael T. Nickerson

Abstract

The effect of polyethylene glycol (PEG) 3350 addition (3%, flour wt. basis) on the properties of dough made from two Canadian Western Red Spring wheat cultivars (*Triticum aestivum* L. 'Harvest' and 'Pembina') differing in dough mixing requirements and dough-handling properties was investigated in a low salt dough formulation (1% NaCl, flour wt. basis). PEG was added for experimental purposes to alter water mobility to better understand underlining mechanisms, however would not be used in real bread formulations. For cultivar Harvest, but not Pembina, dough stickiness was reduced by the addition of PEG. Dough freezable water content decreased with the addition of PEG for both cultivars. Rheological measurements showed that PEG increased dough stiffness as measured by the complex modulus $|G^*|$. Creep measurements indicated that the relative elastic component (J_{el}) increased whereas maximum deformation (J_{max}) decreased with the addition of PEG for cultivar Harvest only. Dough made with a weaker cultivar (Harvest) with the addition of PEG performed similarly to dough made with a stronger cultivar (Pembina) without PEG. Results indicate that in a low sodium environment, availability of water is critically important for controlling a number of properties that relate closely to dough machinability, especially in a weaker wheat cultivar.

Keywords: dough stickiness, polyethylene glycol 3350, dough rheology, wheat cultivar, sodium chloride

1. Introduction

Wheat flour dough in its most basic form (flour and water) is a complex diverse system comprised of starch and non-starch polysaccharides, gluten proteins, enzymes, etc. Dough components can be solubilized or adsorbed using different amounts of water as the water moves between the different phases during mixing. Water becomes either bound or remains free within the dough and this, along with the initial quantity of water added, affects the rheological properties of the dough which in turn governs its machinability [1, 2]. Reducing salt in bread is a major goal in global salt reduction strategies due to the ubiquitous nature of bread in the diet. Salt is a necessary ingredient in large scale bread production contributing to its flavor, and more importantly, to the physicochemical properties of the dough.

IntechOpen

Sodium and chloride ions increase the ordering of water structure which promotes protein–protein hydrophobic interactions. These hydrophobic interactions are the major factor involved in the formation of the gluten network which imparts the viscoelastic dough properties required for baking [3, 4]. Without the addition of salt, dough becomes sticky when mixed due to the over hydration of the gluten proteins [5]; stickiness is a major limiting factor of salt reduction. A sticky dough will cause problems in a bakery throughout the production process, from dough sticking to mixing equipment to a decrease in bread yield [6, 7].

Osmotic regulators such as polyethylene glycol (PEG) represent an experimentally applicable means of altering water availability to gluten to better understand the underlining mechanisms involved within a low sodium environment. However, it would not be used in real dough/baking applications since its known to cause nausea, abdominal pain cramping and gas, and/or cause allergic reactions upon consumption. PEG is a water soluble non-ionic surfactant, which contains the repeat unit (–O–CH_2–CH_2–), where the hydrophilic oxygen are separated by the hydrophobic ethylene units, with a terminal end of H or OH [8]. In previous work Yovchev et al. [9] reported PEG 400 to be a dough softener however no significant differences were found between PEG of different molar masses (PEG 400, ~1600, 3350) on dough rheology or freezable water content when added at a concentration of 1% (flour weight basis). Preliminary work (data not shown) indicated that when the PEG addition level was increased from 1 to 3%, doughs with PEG 3350 behaved much differently than those with the lower molecular weight PEG (PEG 400). Based on this and the previous study [9] the amount of PEG 3350 was increased to 3% and the objective of this study was to investigate how this desiccant would affect dough rheology (stickiness and mechanical properties) and dough water distribution (freezable water content) in low salt (1% NaCl, flour weight basis) formulations as a tool to better understand the impact of water mobility in dough handling.

2. Materials and methods

2.1 Dough formulation and preparation

Two Canadian Western Red Spring (CWRS) wheat cultivars (*Triticum aestivum* L. ‘Harvest’ and ‘Pembina’) were used in this study. Historical breeder data was used in order to select a strong (Pembina) and weak (Harvest) cultivar based on gluten strength. The dough formulation consisted of 10 g flour (14% moisture basis), 0.1 g salt (NaCl), optimal water content and either with or without PEG 3350 (0.3 g). Optimal water absorption was determined from the Farinograph absorption of each cultivar (64.9% for Harvest and 62.0% for Pembina). Dough samples were mixed to peak development using a 10 g mixograph (TMCO National Mfg., Lincoln, NE).

2.2 Stickiness measurements

Dough stickiness measurements were performed using a TA.XTPlus texture analyzer with a Chen-Hoseney dough stickiness cell [10], a 25 mm perspex cylinder probe and 5 kg load cell (Texture Technologies, Stable Micro Systems, Ltd., Surrey, UK). The compression force was 40 gf, pre-test and test speed 0.5 mm/s, post-test speed 10 mm/s, return distance 15 mm, contact time 0.1 s, trigger force 5 gf, and a force vs. time curve was generated. Dough stickiness was taken as the maximum positive force (gf) of the curve. Measurements were performed in duplicate.

2.3 Differential scanning calorimetry (DSC)

Freezable water content (FWC) was measured with DSC Q2000 (TA Instruments, New Castle, DE, USA) containing a refrigerated cooling system. A small piece of dough (~15 mg) was weighed into an aluminum DSC pan then hermetically sealed. An empty pan was used as reference. The pans were equilibrated at 30°C for 5 min then cooled and held at −40°C for 5 min followed by heating to 40°C. The cooling/heating rate was 10°C/min. The freezable water content was calculated directly from the enthalpy (ΔH) of melting peak divided by the enthalpy of pure water, and expressed per gram dry matter. Measurements were performed in triplicate.

2.4 Dough rheology

Dough rheological properties were determined using an AR-1000 rheometer (TA Instruments, New Castle, DE, USA), with a 40 mm parallel plate. A 10 min rest period was employed after the dough was positioned between the plates and the gap was set to 2 mm. An oscillatory frequency sweep from 0.1 to 100.0 Hz was performed at a constant strain of 0.1% (within the linear viscoelastic region). Immediately after the frequency sweep, a creep recovery protocol was carried out on the same dough sample according to the method of Jekle and Becker [11]. For 180 s a shear stress of 250 Pa (τ0) was applied then removed and relaxation was recorded for 360 s to reach a steady state of recovery. Strain was recorded as a function of time and data was collected based on compliance using Eq. (1):

$$J(t) = \gamma(t)\,\tau_0^{-1} \tag{1}$$

where J is the compliance, γ is the strain, and τ_0 is the constant stress which was applied during the creep phase. The creep compliance J_{max} is the shear deformation at t = 180 s of the creep phase. The creep recovery compliance J_r (at t = 360 s of the recovery phase) is a measure of the mechanical energy stored in the sample during the creep phase. The relative elastic part J_{el} [−] was reported using Eq. (2):

$$J_{el} = J_r(J_{max})^{-1} \tag{2}$$

All measurements were performed at a constant temperature of 30°C. Measurements were performed in triplicate.

2.5 Statistical analysis

A one way analysis of variance was performed with a Tukey test to measure differences among means using SigmaStat 4.0 (Systat Software Inc., San Jose, CA, USA).

3. Results and discussion

3.1 Dough stickiness

The stickiness (gf) values for cultivars Harvest and Pembina with and without the addition of PEG 3350 are reported in **Table 1**. Cultivar Harvest produced a much stickier dough than Pembina (86.5 vs. 47.5 gf). Under low salt conditions the addition of PEG significantly decreased dough stickiness (86.5 to 56.3 gf) for cultivar Harvest, whereas for Pembina the stickiness remained unchanged. The addition of PEG induced a greater decrease in stickiness for cultivar Harvest than for Pembina which was most likely due to the higher starting stickiness value of Harvest. With the addition of PEG, a wheat cultivar prone to stickiness (Harvest) could behave more like a stronger wheat cultivar (Pembina) not containing PEG which is hypothesized to be due to the uptake of excess water by PEG in the dough made with Harvest resulting in better dough machinability. It is known that an increase in water content of a dough increases dough stickiness [11, 12].

3.2 Freezable water content

The freezable water content of the doughs with and without PEG was studied to find that the addition of PEG decreased the freezable water content in the dough for both cultivars (**Table 1**). The decrease was approximately the same magnitude for both cultivars; 0.50 to 0.42 g ice/g db for cultivar Harvest and 0.45 to 0.37 g ice/g db for Pembina. This decrease in the freezable water relates to more water molecules being bound in the dough and less free water in the system. It is hypothesized that the water molecules are bound to PEG and constrained in their mobility and as a result, gluten is less hydrated. Doughs made with cultivar Harvest contained more unbound water than dough made with Pembina as seen from the higher freezable water values for Harvest; this causes Harvest to produce stickier doughs than Pembina. The DSC results of PEG decreasing freezable water correspond to the texture analyzer results of PEG decreasing dough stickiness values in cultivar Harvest since free water in the dough contributes to dough stickiness [12, 13].

3.3 Rheological properties

The oscillatory shear measurements and creep recovery results are reported in **Table 1**. Overall the rheological measurements demonstrate that cultivar Pembina produces a stronger gluten network than Harvest. The complex shear modulus |G*| recorded at 1 Hz increased with the addition of PEG for both cultivars, but

	Harvest		Pembina	
	No PEG	With PEG	No PEG	With PEG
Stickiness (gf)	86.5 ± 9.2^{a}	56.3 ± 0.7^{b}	47.5 ± 1.4^{b}	44.7 ± 1.5^{b}
Freezable water (g ice/g db)	0.50 ± 0.01^{a}	0.42 ± 0.01^{b}	0.45 ± 0.02^{c}	0.37 ± 0.00^{d}
G' (10^{3}Pa)	5.5 ± 0.2^{a}	9.2 ± 0.2^{b}	8.3 ± 0.3^{b}	10.5 ± 0.6^{c}
G" (10^{3}Pa)	2.2 ± 0.1^{a}	4.0 ± 0.1^{b}	3.1 ± 0.1^{c}	4.4 ± 0.2^{d}
|G*| (10^{3} Pa)	5.9 ± 0.2^{a}	10.0 ± 0.3^{b}	8.9 ± 0.3^{c}	11.3 ± 0.6^{d}
Tan δ	0.41 ± 0.01^{a}	0.44 ± 0.00^{b}	0.38 ± 0.00^{c}	0.42 ± 0.00^{d}
J_{el}	0.46 ± 0.03^{a}	0.66 ± 0.01^{b}	0.68 ± 0.01^{bc}	0.71 ± 0.01^{c}
J_{max} (10^{-3} Pa^{-1})	4.36 ± 0.28^{a}	2.00 ± 0.01^{b}	1.64 ± 0.00^{bc}	1.49 ± 0.13^{c}

Values are reported as the mean ± standard deviation.
Means in each row followed by different letters are significantly different ($p < 0.05$).
Oscillatory frequency sweep measurements (G', G", |G|, and tan δ) at 1 Hz.*

Table 1.
Stickiness values, freezable water content, and rheological properties of wheat doughs with and without 3% (flour weight basis) PEG 3350.

the magnitude of increase was much greater for Harvest (5.9–10.0 10^3 Pa) than for Pembina (8.9–11.4 10^3 Pa), which was due to the stronger gluten network formed by Pembina without PEG. Cultivar Pembina with PEG produced the stiffest dough with the highest |G*| value out of the four doughs whereas Harvest without PEG had the lowest |G*| value. Despite this stiffening of dough by PEG, it does not improve the gluten network since the tan δ (taken at 1 Hz) of both cultivars increased. PEG addition increased both the loss (viscous) and storage (elastic) modulus of the dough, however the magnitude of increase was larger in the loss modulus therefore the tan δ (G"/G') increased (**Table 1**). It is hypothesized that the PEG is concentrating the gluten network and dough polymers making it stronger, and therefore increasing the G' values, while also having an independent plasticizing effect on the dough which increases the G" values. At a lower molecular weight PEG is predominantly a dough plasticizer [9]. Forces between PEG and proteins are primarily repulsive [14]. From the creep recovery test the relative elastic part, J_{el}, and maximum deformation, J_{max}, were reported. The relative elastic part increased with the addition of PEG for cultivar Harvest to a comparable level with Pembina without PEG, whereas PEG had no significant effect on the dough elasticity as measured by J_{el} for Pembina. The maximum deformation of the dough was greatly reduced for doughs made with cultivar Harvest containing PEG; the J_{max} decreased from 4.36 to 2.00 10^{-3} Pa^{-1}. There was a significant difference between the two cultivars used with Pembina deforming less than Harvest, however Harvest with PEG (J_{max} 2.00 10^{-3} Pa^{-1}) deformed to a similar level as Pembina without PEG (J_{max} 1.64 10^{-3} Pa^{-1}). Therefore, PEG is substantially increasing the short-term relaxation of gluten in the weak cultivar, but because PEG effectively ties up water molecules, water is unavailable to facilitate large deformation of either dough over a longer timescale.

4. Conclusions

The interactions between water and wheat flour components are responsible for dough machinability. The addition of PEG 3350 to a low salt dough formulation improved dough machinability of a weaker flour cultivar (Harvest) by tying up the water. For cultivar Harvest the addition of PEG 3350 decreased dough stickiness, increased dough stiffness (|G*|) and dough elasticity (J_{el}), and decreased dough deformation (J_{max}). All of the aforementioned measurements for Harvest with PEG gave values similar to the stronger wheat cultivar, Pembina, without the addition of PEG. The addition of PEG 3350 also decreased the freezable water content of the doughs made from either cultivar indicating that PEG was removing free or loosely bound water from the dough polymer network. The use of a strong osmotic regulator, such as PEG 3350, in the present experiments, highlights the role of water mobility in governing dough stickiness. It is important to note, PEG 3350 was added for experimental purposes only, and should not be included in real dough/baking applications because of health concerns.

Acknowledgements

Financial support for this research was provided by the Saskatchewan Agricultural Development Fund, the Western Grains Research Foundation, the Natural Sciences and Engineering Research Council of Canada- Collaborative Research Development Program, the Canadian Institute of Health Research, and Canada Bread Corp. (Toronto, Canada).

Conflict of interest

The authors declare no conflict of interest.

Author details

Andrea K. Stone[1], Aleksandar Yovchev[1], Pierre J. Hucl[2], Martin G. Scanlon[3] and Michael T. Nickerson[1*]

1 Department of Food and Bioproduct Sciences, University of Saskatchewan, Saskatoon, SK, Canada

2 Crop Development Centre, University of Saskatchewan, Saskatoon, SK, Canada

3 Department of Food and Human Nutritional Sciences, University of Manitoba, Winnipeg, MB, Canada

*Address all correspondence to: michael.nickerson@usask.ca

References

[1] Assifaoui A, Champion D, Chiotelli E, Verel A. Rheological behaviour of biscuit dough in relation to water mobility. International Journal of Food Science and Technology. 2006;**41**:124-128. DOI: 10.1111/j.1365-2621.2006.01469.x

[2] Cherian G, Chinachoti P. ^{2}H and ^{17}O nuclear magnetic resonance study of water in gluten in the glassy and rubbery state. Cereal Chemistry. 1996;**73**:618-624

[3] Danno G, Hoseney RC. Effect of sodium chloride and sodium dodecyl sulfate on mixograph properties. Cereal Chemistry. 1982;**59**:202-204

[4] Hlynka L. Influence of temperature, speed of mixing and salt on some rheological properties of dough in the farinograph. Cereal Chemistry. 1962;**39**:286-303

[5] van Velzen EJJ, van Duynhoven JPM, Pudney P, Weegels PL, van der Maas JH. Factors associated with dough stickiness as sensed by attenuated total reflectance infrared spectroscopy. Cereal Chemistry. 2003;**80**:378-382. DOI: 10.1094/CCHEM.2003.80.4.378

[6] Adhikari B, Howes T, Bhandari BR, Truong V. Stickiness in foods: A review of mechanisms and test methods. International Journal of Food Properties. 2001;**4**:1-33. DOI: 10.1081/JFP-100002186

[7] Huang WN, Hoseney RC. Isolation and identification of a wheat flour compound causing sticky dough. Cereal Chemistry. 1999;**76**:276-281. DOI: 10.1094/CCHEM.1999.76.2.276

[8] Crupi V, Jannelli M, Magazu S, Maisano G, Majolino D, Migliardo P, et al. Neutron scattering and compressibility measurements for the study of hydration effects on polymeric aqueous solutions. Il Nuovo Cimento D. 1994;**16**:809-816. DOI: 10.1007/BF02456728

[9] Yovchev AG, Stone AK, Hucl P, Scanlon MG, Nickerson MT. Effects of salt, polyethylene glycol, and water content on dough rheology for two red spring wheat varieties. Cereal Chemistry. 2017;**64**:513-518. DOI: 10.1094/CCHEM-09-16-0240-R

[10] Chen WZ, Hoseney RC. Development of an objective method for dough stickiness. LWT- Food Science and Technology. 1995;**28**:67-473. DOI: 10.1006/fstl.1995.0079

[11] Jekle M, Becker T. Dough microstructure: Novel analysis by quantification using confocal laser scanning microscopy. Food Research International. 2011;**44**:984-991. DOI: 10.1016/j.foodres.2011.02.036

[12] Jekle M, Becker T. Effect of acidification, sodium chloride, and moisture levels on wheat dough: I. modeling of rheological and microstructural properties. Food Biophysics. 2011;**7**:190-199. DOI: 10.1007/s11483-012-9257-0

[13] Beck M, Jekle M, Becker T. Impact of sodium chloride on wheat flour dough for yeast-leavened products. I. Rheological parameters. Journal of the Science of Food and Agriculture. 2012;**92**:585-592. DOI: 10.1002/jsfa.4612

[14] Bhat R, Timasheff S. Steric exclusion is the principal source of the preferential hydration of proteins in the presence of polyethylene glycols. Protein Science. 1992;**1**:1133-1143. DOI: 10.1002/pro.5560010907

Chapter 11

Glycation of Animal Proteins Via Maillard Reaction and Their Bioactivity

Blanca Areli Mondaca-Navarro, Roberto Rodríguez Ramírez, Alma Guadalupe Villa Lerma, Luz Angelica Ávila Villa and Gabriel Davidov Pardo

Abstract

Nowadays there has been an increase in the need to incorporate foods in our diets that have optimal and palatable organoleptic characteristics as well as complex interaction in human biological processes that provide beneficial properties to human health. Animal foods and their by-products are an important source of macro- and micronutrients and also a great protein source; nevertheless the consumption of these products has been decreasing since they have been associated with the generation of chronic degenerative diseases; therefore the food industry has sought to innovate toward the generation of healthier foods. This chapter presents an overview of the glycation of proteins of animal origin via the Maillard reaction emphasizing on their posttranslational modifications and their possible uses in food, based on their bioactivity.

Keywords: glycation, Maillard reaction, bioactivity

1. Introduction

Nutritional and functional characteristics of proteins have been a subject of research for many years since they own catalytic, regulatory, and structural functions which are fundamental for several biological processes occurring in living beings. These macromolecules are formed by a combination of α-amino acids that are widely distributed in nature both from plant (legumes, cereal grains, nuts, fruits, and vegetables) and animal origin (dairy, meat, seafood, eggs) [1]. Each protein structure has a unique chain of amino acids linked together by a peptide bond providing bioactivities. The nutritional quality of a protein depends on its amino acid content and the physiological use of specific amino acids after digestion and absorption.

On the other hand, proteins give functional (structural-physicochemical) and bioactive (nutritional-health) properties to food products. These properties modify certain characteristics of foods that are industrially processed or used in households. Proteins modify the sensory properties as well as rheological behavior, stability, and nutritional value of such foods [2–4].

Previous research has questioned the safety of proteins that have been structurally modified to improve their functionality and bioactivity based on the dynamics of their environment [5]. In some case these modifications have resulted in the presence of genetic mutations, variation in the structure of the protein, as well as the generation of posttranslational modifications [6].

The importance of proteins in the health of humans is an opportunity for the development of new protein ingredients toward the improvement of separation processes and the exploitation of proteins obtained from different by-products considered as waste. During these processes, it is important to consider the functional properties of proteins in foods as well as their transformation during food processing and the metabolic behavior on the digestibility and bioavailability during gastrointestinal digestion [1].

Maillard reaction is one of the most common and important phenomena that takes place during the thermal processing both at an industrial and household levels or even during the storage of foods with high contents of reducing sugars and proteins [7]. Considering the fact that Maillard reaction occurs in several foods of common consumption and that it is generated spontaneously in foods under heat treatments, it is important to determine its safety and influence on the nutritional value of food products. Despite the ability of the Maillard reaction to affect color, taste, and texture of most foods, the compounds generated at different stages of this reaction can exert effects on the human body. In the past, Maillard reaction was mostly investigated for its negative effects, such as the loss of the nutritional value of the food. This is mainly attributed to inactivation or destruction of amino acids, decrease in the digestibility of nitrogen, and impaired absorption of brown compounds (melanoidins) in the intestine [8].

On the basis of this context, it is intended to show that, during the modification of proteins via Maillard reaction, the compounds that are formed do not have adverse effects, but actually have positive effects related to the bioactivity that they offer. Due to this reason in recent years investigations have been focused on the biological functionality of glycated protein resulting from the Maillard reaction from the point of view of the development of new foods or food additives.

2. Posttranslational modifications of proteins

Posttranslational modifications in proteins are defined as "covalent processing events that change the properties of a protein by proteolytic cleavage or by adding a modifying group to one or more amino acids" [9]. Knowing and analyzing these modifications of food proteins represents a challenge to understand their technological and biological function.

It is possible to find posttranslational modifications in animal proteins due to the action of different physical, enzymatic, and chemical treatments (acetylation, deamination, nitration, methylation, lipidation, carboxylation, formation of disulfide bonds, hydroxylation, sulfation, amidation, and glycosylation) [10–12]. A wide variety of posttranslational modifications have been characterized; some of them are formed mostly in intracellular proteins such as the phosphorylation mechanism, and conversely there are some other processes such as glycosylation, nonenzymatic glycation, formation of disulfide bonds, and carboxylation, which are formed in extracellular proteins [13].

Depending on the reaction mechanisms applied to them, protein modifications originate a wide range of characteristics (reaction, location, transformation) that could be studied through the science and technology of food products of animal origin. However, there is a limitation when using chemicals for the formation of posttranslational modifications in proteins, since these could induce the production of toxic compounds resulting in the development of unfit food for human consumption [4, 13–15].

3. Maillard reaction

Maillard reaction is one of the most common, spontaneous, and important actions that are formed in the processing and storage of food. This set of chemical reactions give rise to the formation of brown pigments, odors, and flavors, as well as to alterations in the functionality, nutritional value, and shelf life of protein-rich foods. This reaction is also known as a nonenzymatic process that arises from the heat-catalyzed covalent condensation of a carbonyl-containing compound and a deprotonated amino group [7, 16]. This reaction was first discovered by Louis Camille Maillard in 1912, who established that this reaction alters the nutrients in food during cooking [17]. The first findings obtained from the Maillard reaction are briefly described below [18].

- Maillard reaction is universal, regardless of the nature of the amino group or the aldehyde/ketone group corresponding to reducing sugar.
- The reaction applies to nucleic acids, amino acids, peptides, and proteins.
- The aldehyde or the ketone group of sugar is essential for the reaction to take place.
- High temperatures are not essential. The reaction can be slowly carried out at 34°C and even at 15°C, as long as the mixture has been previously heated.
- Oxygen does not interfere with the reaction.
- During the reaction, the early products undergo extensive dehydration.
- In Maillard reaction there is a release of carbon dioxide, which comes from the amino acid.

3.1 Maillard reaction formation

Recently in 2017, Taghavi et al. [19] described Maillard reaction, as a series of complex reactions. Despite the advances in science and research related to this subject, Maillard reaction mechanism is not fully known, due to the reactivity and complexity of it. According to scientific literature, the most accepted route used as a reference to understand Maillard reaction is the one proposed by Hodge [20] which divided the process into three stages with seven different reactions (**Figure 1**):

1. Initial stage

    A. Condensation: Schiff base formation

    B. Amadori rearrangement

2. Intermediate stage

    A. Dehydration of sugars

B. Fragmentation of sugars

C. Degradation of amino acids

3. Final stage

A. Polymerization and aldol condensation

B. Formation of heterocyclic nitrogen compounds and colored products

3.2 Description of Maillard reaction stages

Initial stage: Condensation (A) and rearrangement of Amadori or Heyns (B). At this stage the reaction consists of a simple condensation between a carbonyl group of a reducing sugar and a free amino group of an amino acid, protein, nucleic acid, or a low molecular weight amine, producing an N-substituted glycosylamine also called Schiff base. This step is considered a reversible reaction since in a strong acidic medium it can be protonated into the carbonyl of the reducing sugar and the free amino group from the amino acid. Schiff base is cyclized to form an N-substituted glycosylamine; an aldosamine or ketosamine can be formed depending on the type of reducing sugar that took part in the reaction, an aldose or a ketose, respectively. The molecule then forms an N-substituted-aldosamine, and through the Amadori rearrangement, a 1-amine-1-deoxy-2-ketose is formed. However, when the molecule is an N-substituted-ketosamine, a 2-amine-2-deoxy-2-aldose molecule is formed by means of the Heyns rearrangement [4].

The formation of Schiff bases and of the Amadori/Heyns rearrangements are the Maillard reaction stages where more research has been focused on; thus their structures have been precisely determined, as well as the properties of the first products

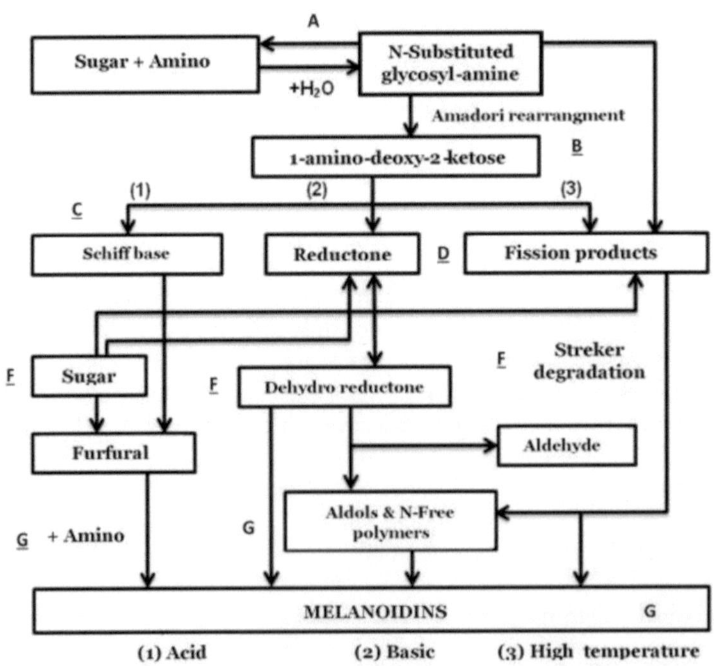

Figure 1.
Maillard reaction depiction adapted from reference [20].

involved in this stage [21]. Amadori and Heyns products decompose depending on the pH and the temperature of the medium, resulting in the formation of different intermediate compounds. Specifically, the conditions or physical variables involved in the reaction are temperature and heating time, pH, water activity (Aw), as well as the type and availability of the reactants (amino acid and reducing sugar) [22], which directly affects the formation of glycoconjugates that are not naturally found in foods [23].

The intermediate stage includes dehydration of sugars (C), sugar fragmentation (D), and Strecker degradation (E) where a 1,2-enolization is formed at low pH, giving rise to the formation of dicarbonyl (precursors of brown compounds) and finally furfural or hydroxymethylfurfural (HMF). On the other hand, at basic pH 2,3-enolization is formed, and the final compounds are reductones, which can be dehydrated to form dehydro-reductones, forming polymers when reacting with amino groups [24]. Amadori compounds can be split into different products of low molecular weight such as glyceraldehyde, pyruvaldehyde, acetol, acetoin, and diacetyl [25]. All of them having a characteristic odor and high reactivity and being unsaturated substances, they follow various chemical routes depending on the pH, temperature, and Aw conditions [26].

Final stage: Hodge [27] described this stage as a sequence of reactions such as aldol condensation and polymerization (F) and formation of heterocyclic nitrogen compounds and colored products (G). These joint reactions show that amino compounds are effective catalysts for aldol condensation and polymerization, as well as oxidation and other spontaneous reactions, to produce a series of chemical compounds that are known as advanced glycation end products (AGEs). On the other hand, formation of nonvolatile colored and high molecular weight nitrogen compounds called melanoidins [28–30] is also carried out during the final stage. The structure of melanoidins is primarily made of aldehyde groups due to the degradation of sugars formed in the intermediate stage of Maillard reaction which are polymerized and bound to amino groups [31]. It is complicated to explain the progress of Maillard reaction in foods, due to the presence of different reactive groups mixed together (reducing amino acids and sugars) and to the dynamic conditions of food matrices that favor the formation of polymers. All this contributes to the difficulty of fully characterizing the later stages of the reaction due to their complicated chemical perspective [32].

4. Protein glycation via Maillard reaction

One of the technologies for the formation of posttranslational modifications in food proteins is nonenzymatic glycation, also known as glycation via Maillard reaction or simply glycation. This process is known as the chemical interaction between a reducing sugar with an amino group of peptides and proteins forming a covalent bond and leading to the formation of glycated proteins or glycoconjugates. Compared to other chemical methods used for protein modification, Maillard reaction glycation does not imply the use of reagents that may affect human health, which results in the creation of proteins with new technological and biological interest [2, 14].

In the early twentieth century, there was a very limited knowledge about glycoproteins, these were described as "Compounds of the protein molecule mixed with a substance containing a carbohydrate group that is not a nucleic acid" claiming that glycoproteins were those proteins containing a glycosyl bond [33]. However, years later the term glycation was defined as "All reactions that bind a sugar to a protein or a peptide, whether catalyzed or not by an enzyme" [19]; otherwise Lis and Sharon in 1993 [34] theoretically differentiated the term enzymatic glycosylation and nonenzymatic glycation of sugar-bound proteins.

Regarding the process of modification of proteins by means of enzymatic glycosylation, there is the intervention of glycosyltransferases and nucleotides as sugar donors which form glycoproteins [35]. In the development of glycated proteins, nonreducing sugars cannot be used since the interaction of their aldehyde or ketone groups is not possible [34, 36, 37]. Glycoconjugates have the uniqueness of possessing bioactivity and functionalities that act to improve the native protein. Research have been focusing on the toxic effect of Maillard reaction on proteins due to the formation of the different intermediary compounds, which in consequence form amino acid derivatives (AGEs) [38–41]. It is convenient for the formation of glycoconjugates to avoid the advanced stages of the reaction, in the same way it is important to understand the complexity of the Maillard reaction for the development of new food products with added value [2, 4].

5. Bioactivity in glycoproteins of animal origin obtained via Maillard reaction

In recent years, and due to current global trends, consumers have demonstrated a potential interest and awareness about the importance of correlating health with food and achieving the combination of nutritional value, bioactivity, and improved organoleptic properties to those already existing in natural and processed foods, as well as their interaction with biological systems, which help to prevent pathological conditions. These changes in the food standards of consumers generate a field of opportunity for the formulation of new foods, so the search for bioactive properties in proteins of animal origin has been a challenge for food scientists and technologists, where research has been focused on the glycation of proteins with interest to the food industry.

5.1 Antioxidant capacity

The concept of antioxidant is defined as "any substance that, in presence of an oxidizable substrate, significantly delays or inhibits the oxidation of that substrate" [42–44]. The concept of antioxidant was first originated from chemistry and then adapted to biology, medicine, epidemiology, and nutrition. It describes the redox ability of molecules in food and biological systems and their interaction with free radicals [45].

In recent decades there has been a growing interest in the study of the antioxidant activity of food and diets of humans. The deficit of antioxidants in the body causes negative biological effects that lead to the presence of free radicals and as a consequence to the development and progress of chronic degenerative diseases related to oxidative stress. Based on scientific literature, oxidative stress is characterized by an increase in free radicals of reactive oxygen and nitrogen species, as well as by a decrease in the body's antioxidant defenses [46]. This effect is associated with bad eating habits, generating an imbalance between antioxidant systems and the production of oxidizing agents that lead to the generation of several chronic degenerative pathologies that attain human health [47, 48].

Antioxidants in conjunction with the daily diet have great relevance since they have a determining role in human health, which help to reduce oxidative stress. In dietary matrices antioxidants are useful in delaying lipid peroxidation and therefore help in taste, texture, and in some cases color [49]. In the human body endogenous antioxidants help to protect, prevent, or delay cell death, tissue and organs lesions, as well as oxidative damage by reactive oxygen and nitrogen species [50].

Some of these studies are focused on giving an added value to the by-products of food industry, focusing on the analysis of antioxidant activity; an example of this is the use of pork blood as a protein source, which is an abundant by-product in the slaughter process. Porcine blood plasma contains a variety of bioactive compounds and high-quality proteins; for this reason there are several studies on the antioxidant capacity of glycoconjugates formed from proteins of pork serum with glucose, fructose, and galactose, which were glycated via Maillard reaction at 100°C, reporting that glycoconjugates formed with glucose showed the lowest antioxidant capacity, while the glycoconjugates formed with galactose were the ones showing the highest antioxidant capacity [51].

Following with the investigation on pork blood plasma proteins, different researchers determined the antioxidant capacity in glycoconjugates formed with 2% blood plasma and 2% glucose adjusted to several pH (8, 9, 10, 11, and 12) and 100°C at different heating times (0–8 h) reporting that glycoconjugates formed at higher pH (pH 12) showed a higher antioxidant activity than the ones obtained at lower pH levels, showing that pH was a factor in determining the antioxidant activity of glycoconjugates [52]. Years later, researchers reported that the glycation of hydrolyzed porcine serum plasma with glucose, fructose, or galactose heated at 95°C, for up to 6 h, yielded an antioxidant capacity of 45% [3].

A vast amount of research have been carried out on the milk protein glycation and its antioxidant action analyzed by different methodologies from which ABTS and DPPH techniques stand out. Most of the research have focused on dairy proteins such as lactoglobulins, caseins, and whey proteins. As an example glycoconjugates formed with hydrolyzed β-lactoglobulin with glucose heated to 90°C for a maximum of 18 h showed an antioxidant activity greater than 50% [53].

In another research Stanic-Vucinic et al. [54] tested high-intensity ultrasound in order to promote glycation of β-lactoglobulins with glucose, galactose, lactose, fructose, ribose, and arabinose via Maillard reaction at pH 6.5. As a result, an increased activity of DPPH radical inhibition was reported, with the glycoconjugates being the ones that showed the highest antioxidant activity with 42%. Similarly, glycoconjugates with ribose and milk proteins such as α-lactalbumin or β-lactoglobulin, heated at 95°C, for up to 5 h exhibited an increase in their antioxidant activity [55].

Different researchers formed glycoconjugates using casein and glucose, at pH 12 and at a temperature of 102°C for 130 min. Glycoconjugates were ultrafiltered in order to form fractions of different molecular weights (50 kDa, 30 kDa, 10 kDa, 5 kDa, 1 kDa) and to analyze the antioxidant activity by the DPPH method. It was shown that glycoconjugates with higher molecular weights were the ones with the best in vitro radical scavenging activity [56]. Jing and Kitts [57] formed glycoconjugates with glucose, fructose, and ribose at 55°C, pH 7.0 for 28 days. A concentration range of 0.1–0.5 mg/ml of sugar and casein was used, indicating that only the casein-ribose system showed antioxidant capacity with the DPPH radical inhibition technique (3–7%). On the other hand glucose-casein and fructose-casein systems showed no antioxidant activity in the concentrations and treatments tested.

There are only a few works related to glycated proteins of marine origin and their relationship with antioxidant capacity. Decourcelle et al. [58] evaluated the formation of glycoconjugates at 50°C for 48 hours between a shrimp hydrolyzate with xylose or dextran, reporting that the antioxidant capacity of the shrimp hydrolyzate mixed with xylose was 13.5 to 16 times higher than the shrimp hydrolyzate without the mixed carbohydrate.

Different researchers affirm that glycation via Maillard reaction could cause structural changes in protein molecules, which could generate a wide range of compounds or Maillard reaction products (MRP), which would lead to the formation of conjugates and that these could contribute to the generation or increase of antioxidant capacity [59].

5.2 Chelating capacity

Transition metals are a systemic and determining part for the proper functioning of the structural and functional components of the organism; these metals have the ability to donate and accept electrons. Although they are also limited by the fact that transition metals can lead to toxicity that occurs when one or more of these metals are increased in the body, moreover the presence of transition metals favor the formation of free radicals [60]. Chelating agents are compounds capable of binding to metal ions leading to the formation of chelates; this metal-ion complex can carry a positive, negative charge, or in turn no charge [61].

Chelating agents have a "ligand" which binds the atoms that form the chelate with either two covalent bonds or a covalent bond and a coordinate bond or two coordinated bonds when bidentate chelates are formed. In biological systems the metal ions Na^{+}, Mg^{2+}, Cu^{2+}, Zn^{2+}, and especially transition metals such as Mn^{2+}, Fe^{2+}, and Co are involved in the formation of chelates [62]. Of the aforementioned transition metals, iron and copper, despite being essential micronutrients for optimal organism functioning, have the peculiarity of transferring and gaining electrons from oxygen molecules and forming reactive oxygen species (ROS), which can quickly catalyze the superoxide radical and could generate oxidative stress [60].

There is scientific evidence proving that the molecules formed through the Maillard reaction have the ability to chelate metal ions giving them the characteristic of transforming metal behavior into physiological functions. Irving and Williams [63] were possibly the first researchers to confirm the formation of complexes involving the Maillard reaction and certain transition metal ions and describe the strength bond of chelates formed with metal ions following the order of $Mn^{2+} < Fe^{2+} < Ni^{2+} < Cu^{2+} > Zn^{2+}$. Years later they confirmed the formation of these chelates in an order of $Mg^{2+} > Cu^{2+} = Ca^{2+} > Zn^{2+}$ [64]. This last order differs from that found in 1953, so they conclude that there could be due to the presence of more than one type of ligand [63].

References from previous research provide us with relevant information on the ability to chelate transition metals through glycated proteins via Maillard reaction. Gu et al. [56] formed casein glycoconjugates with glucose affirming that the glycoconjugates with greater molecular weight have greater chelating potential of ion Fe^{+2} than those of low molecular weight. Different researchers reported the formation of glycoconjugates with β-lactoglobulin hydrolyzate obtaining 64 and 61% iron chelation, for the glycoconjugate of β-lactoglobulin-glucose and β-lactoglobulin hydrolyzate, respectively. From the obtained results, they concluded that the increase in iron chelating activity was mainly related to the glucose caramelization reaction and not to the glycation process [53].

During the formation of β-lactoglobulin glycoconjugates with glucose, galactose, lactose, fructose, ribose, and arabinose via Maillard reaction, at pH 6.5, it was demonstrated that both native β-lactoglobulin and treated β-lactoglobulin possess chelating activity, in the same way the glycation of β-lactoglobulin in the presence of all sugars resulted in a significant increase in the ability of chelation of iron ($p < 0.05$), being the β-lactoglobulin-ribose system the one with the most prominent effect [54]. In the case of the model system consisting of glucose-asparagine-chitosan with different molecular weights of chitosan and a thermal treatment of 180°C for 30 min, a 60% ferrous ion chelation was obtained in all samples analyzed [65].

You et al. [66] tested the formation of glycoconjugates of silver carp hydrolyzate with glucose (w/w = 2:1, 1:1, 1:2, 1:4) at pH 7.5 and at 50 and 60°C for 24 h reporting that the glycoconjugate with a protein-carbohydrate ratio of 2:1 showed a greater chelating effect (90%) than other ratios tested.

The effect of metal chelation has been related to the hydroxyl and pryoline groups, and also it has been attributed to the steric effects and the multiple interactions of the proteins in the Maillard reaction. However, other authors attribute the ability to chelate Fe^{2+} ions to the higher molecular weight MRPs such as melanoidins and hydroxyl groups formed in the final stage of the Maillard reaction [67, 68].

5.3 Prebiotic and antimicrobial effect

Significant amounts of compounds formed via Maillard reaction are consumed in the human diet, which interact in the body through the digestive system and the intestinal microflora itself; this microflora plays a crucial role for its proper functioning; therefore, maintaining a balance between beneficial and harmful microorganisms is of vital importance. Prebiotics are nondigestible or low digestible food ingredients that benefit the host organism selectively by stimulating the growth or activity of probiotic bacteria in the colon [69].

Different studies have found that products derived from the Maillard reaction can act as prebiotics by activating intestinal microflora, generating a positive effect on *lactobacilli* and *bifidobacteria*. This was observed in a study where intestinal bacteria were able to use the products formed in the different stages of the Maillard reaction, as a substrate, and transform them into energy [70]. Corzo-Martínez et al. [71] glycated dairy proteins specifically β-lactoglobulin and sodium caseinate with galactose and lactose evaluating the effect of hydrolyzed glycoconjugates on the growth of *Lactobacillus*, *Streptococcus*, and *Bifidobacterium* under simulated in vitro gastrointestinal digestion and reported that glycoconjugates formed with caseinate were rapidly fermented by some strains, promoting a greater growth rate than β-lactoglobulin complexes. Therefore, from the results obtained, they could conclude that the conjugation of both dairy proteins with galactose and lactose through the Maillard reaction could be an efficient method to obtain new food ingredients with a potential prebiotic character.

It has also been shown that the compounds formed by the Maillard reaction may be a solution for the infection of pathogenic bacteria such as *Helicobacter pylori*, *Escherichia coli*, *Bacillus cereus*, *Staphylococcus aureus*, *Proteus mirabilis*, *Pseudomonas aeruginosa*, and *Salmonella typhimurium*. The bacteriostatic power of these compounds was studied demonstrating that the action against the bacteria depended on the concentration of reactants, the pH, the temperature, as well as the molecular weight of the glycoconjugates. An example of this is a study performed for the inhibition of *Helicobacter pylori*, where casein-lactose glycoconjugates were given to infected subjects during 8 weeks, obtaining favorable results against *H. pylori* [72]. Similarly, a significant inhibition was observed in mice which received a 10-week treatment with the glycoconjugates [73].

Bacterial adhesion assays have been performed to assess if β-lactoglobulin protein glycated with chitin oligosaccharides (60°C for 6 and 12 h) can be recognized by bacterial adhesins from *Escherichia coli* K88 and *Salmonella choleraesuis*. Biorecognition analyses showed that glycoconjugates were in fact recognized by the adhesins of *E. coli* K88 and *S. choleraesuis*; both adhesins showed an effect similar to mucins (control) which are natural ligands for these microorganisms; that effect was not shown in the native or non-glycated protein concluding that the glycoconjugates formed could be used to investigate the interaction of protein-carbohydrate in biological systems as well as to look for alternatives to bacterial infections [74].

Research on this subject indicate that the antimicrobial activity of the compounds formed by Maillard reaction may be related to their anionic charges and their ability to chelate transition metals such as Fe^{2+}, Zn^{2+}, and Cu^{2+}, which are essential for the proper functioning, survival, and growth of the pathogenic organism [75]. Chung et al. [76] reported the antimicrobial capacity of Maillard reaction products obtained by the condensation of chitosan and glucosamine, against *E. coli* and *Staphylococcus aureus*.

Cell membrane has been analyzed in different bacterial strains of both Gram-positive (*Staphylococcus aureus and Bacillus cereus*) and Gram-negative (*Escherichia coli*, *Proteus mirabilis*, *Pseudomonas aeruginosa*, *and Salmonella typhimurium*) in order to correlate the antimicrobial effect by iron chelation, finding that at low concentrations of melanoidins, a bacteriostatic activity mediated by chelation of iron contained in the culture medium is exerted. Regarding bacterial strains capable of producing siderophores for iron acquisition, the compounds generated from the Maillard reaction chelate the Fe^{3+} siderophore, which could decrease the pathogenicity of the bacteria [67]. It suggests that foods which contain molecules formed by the Maillard reaction may be a beneficial alternative to the intestinal flora and digestive system.

5.4 Antihypertensive capacity

Angiotensin I enzyme (ACE) is also known as peptidyl dipeptidase A or Braquina II; since it eliminates carboxy-terminal dipeptides from a wide variety of peptide substrates [77], it is considered as the mechanism of action of food-derived antihypertensive activity. ACE is essential in the renin-angiotensin system, since it regulates blood pressure and water-salt balance in the human body. The activity of the ACE consists of the conversion of angiotensin I into angiotensin II, which is an effective vasoconstrictor, while degrading bradykinin, which is considered a potent vasodilator [78]. The importance of these mechanisms relies on the blocking of the renin-angiotensin-aldosterone system due to ACE inhibition, which has led to consider ACE inhibitors as first-line therapy against hypertension [79].

Synthetic ACE inhibitors own a mechanism of action consisting in occupying related and specific angiotensin I binding sites. There are commercial drugs such as captopril, enalapril, enalaprilat (active form of enalapril), and lisinopril, capable of inhibiting ACE and therefore decreasing blood pressure levels [80]. Although research has focused on natural origin peptides from different sources to inhibit ACE, it has been shown that some proteins modified via Maillard reaction are capable of presenting antihypertensive activity [81, 82] albeit studies on these kind of proteins is limited, because these research have been focused only on modified peptides [83].

The glycoconjugates formed by casein-xylose (110°C, 30 min) have shown an increase on ACE inhibitory activity when there was a prolonged reaction time of the Maillard reaction, under analyzed conditions [82]. Jiang et al. [84] reported a decrease in antihypertensive activity when analyzing a glycoconjugate formed of a tripeptide with ribose (98°C for a time range 0–8 h) found in the formed glycopeptide [82.] They also formulated glycoconjugates based on casein mixed with different carbohydrates (ribose, galactose, and lactose) showing a decreasing tendency of the ACE inhibitory activity and reporting that this tendency could be attributed to the consumption of casein peptides involved in polymerization during the different stages of the Maillard reaction; although they also confirm that some MRPs are involved in potentiating the ACE inhibitory activity [85]. This is similar to what was reported in a research performed in roasted coffee which demonstrated that melanoidins showed ACE inhibitory activity in vitro [86].

Jiang et al. [87] evaluated the effect of temperature and pH on the inhibitory activity of ACE, forming glycoconjugates with bovine casein and galactose peptides in aqueous solution. Mixtures were heated at 70–120°C for 3 h, at pH 9.0, concluding that as the temperature increased, the ACE inhibitory activity of glycoconjugates gradually decreased.

The possible mechanism of action of ACE against PRM is unknown; however the inhibitory action may be related to chelation of transition metals or its antioxidant activity since ACE depends on metals such as Zn [88].

5.5 Cytotoxic effect

In previous years it was considered that compounds formed during the different stages of Maillard reaction were precursors of carcinogenic activity. This notion was due to the presence of acrylamide or heterocyclic amines that are formed by this reaction, which could have cytotoxic and antiproliferative properties. However, nowadays, there are some investigations that indicate that glycoconjugates obtained via Maillard reaction usually present inhibition of cancer cell proliferation; there it is important to study these compounds since they could bring certain health benefits.

The antimutagenic properties of the Maillard reaction have been widely investigated in in vitro model systems or in complex foods under controlled heat treatment conditions. This bioactive property has been attributed to the inhibition of the absorption of the mutagen or the inhibition of its activation [89]. An example of this has been the study of antiproliferative activity in model systems in 20 amino acids with glucose and fructose, in human colon cancer cells at concentrations of 0.35–1.5 mg/mL. The concentration that showed the highest antiproliferative activity in cancer cells was the fructose-methionine system (32.64%), while for tryptophan it was 15.01%, phenylalanine 30.73%, and tyrosine 21.52%. On the other hand, glycoconjugates derived from glucose mixed with the same amino acids also presented antiproliferative activity [88].

When mixtures of glucose with glycine have been studied, the results were different from those mentioned above since no antiproliferative activity was present [90]. However, when low and high molecular weight melanoidin fractions were isolated, low molecular weight fractions were found to be more reactive to genotoxicity and mutagenicity, although these compounds were considered not to be a health risk [91].

Some glycoproteins may have a non-cytotoxic action; this has been demonstrated through the use of carbohydrate protein model systems [92]. Jing and Kitts [57] tested sugars such as glucose, fructose, and ribose mixed with casein to form glycoproteins (55°C, pH 7, 28 days of incubation) and evaluated their toxicity against Caco-2 cells, concluding that there was no toxicity to the cell model using both low (0.5 mg/mL) and high (2.0 mg/mL) concentrations of analyzed glycoproteins. In agreement with these results, Wei et al. [93] performed an investigation with glycoconjugates formed by bovine albumin serum with galactose, ribose, and lactose showing a low toxicity in the inhibition of Caco-2 cells. Other glycoconjugates formed by ribose-casein peptide and lactose-casein peptides (95°C/5 h) have shown the same results, not presenting cytotoxicity in Caco-2 cells; however another system such as galactose-casein peptides showed a slight decrease in Caco-2 cells [85].

In general and regarding previous research, some of Maillard reaction products that have been synthesized and purified from different protein sources of animal origin can be considered to have an inhibition on the growth of human carcinoma cells in vitro [94].

6. Conclusions

The posttranslational modification of glycated proteins of animal origin has been shown to have or to potentiate specific bioactivity. Therefore, protein glycation of animal origin via Maillard reaction may represent an alternative for the use of by-products of the animal-based food industry. It could be possible to use glycoproteins as a food ingredient, or the incorporation of these bioactives into a technological process in the food industry could result in health benefits, as well as benefits to the environment.

Author details

Blanca Areli Mondaca-Navarro[1], Roberto Rodríguez Ramírez[1*], Alma Guadalupe Villa Lerma[1], Luz Angelica Ávila Villa[2] and Gabriel Davidov Pardo[3]

1 Laboratorio de Biotecnología y Trazabilidad Molecular de los Alimentos, Instituto Tecnológico de Sonora, Ciudad Obregón, Sonora, México

2 Departamento de Ciencias de la Salud, Universidad de Sonora, Ciudad Obregón, Sonora, México

3 Department of Human Nutrition and Food Science, California State Polytechnic University, Pomona, CA, USA

*Address all correspondence to: roberto.rodriguez@itson.edu.mx

References

[1] Sun-Waterhouse D, Zhao M, Waterhouse GI. Protein modification during ingredient preparation and food processing: Approaches to improve food processability and nutrition. Food and Bioprocess Technology. 2014;**7**:1853-1893

[2] Oliver Christine M, Melton Laurence D, Stanley RA. Creating proteins with novel functionality via the Maillard reaction: A review. Critical Reviews in Food Science and Nutrition. 2006;**46**:337-350

[3] Jianhua L, Qiaomei R. Ding Yuting Glycation a promising method for food protein modification: Physicochemical properties and structure, a review. Food Research International. 2012;**49**:170-183

[4] de Oliveira FC, Coimbra JSDR, de Oliveira EB, Zuñiga ADG, Rojas EEG. Food protein-polysaccharide conjugates obtained via the Maillard reaction: A review. Critical Reviews in Food Science and Nutrition. 2016;**56**:1108-1125

[5] Tompa P. Intrinsically disordered proteins: A 10-year recap. Trends in Biochemical Sciences. 2012;**37**:509-516

[6] Broersen K. Technical proteins: A multitude of applications to discover. Journal of Physical Chemistry and Biophysics. 2014;**4**:e123

[7] Mondaca-Navarro BA, Ávila-Villa LA, González-Córdova AF, López-Cervantes J, Sánchez-Machado DI, Campas-Baypoli ON, et al. Antioxidant and chelating capacity of Maillard reaction products in amino acid-sugar model systems: Applications for food processing. Journal of the Science of Food and Agriculture. 2017;**97**:3522-3529

[8] Langner E, Rzeski W. Biological properties of melanoidins: A review. International Journal of Food Properties. 2014;**17**:344-353

[9] Mann M, Jensen ON. Proteomic analysis of post-translational modifications. Nature Biotechnology. 2003;**21**:255-261

[10] Hoffman M, Sniatynsky M, Kast J. Current approaches for global post-translational modification discovery and mass spectrometric analysis. Analytica Chimica Acta. 2008;**627**:50-61

[11] Rathore AS. Follow-on protein products: Scientific issues, developments and challenges. Trends in Biotechnology. 2009;**27**:698-705

[12] Chuh K, Pratt M. Chemical methods for the proteome-wide identification of posttranslationally modified proteins. Current Opinion in Chemical Biology. 2015;**24**:27-37

[13] Walsh G. Post-translational modifications in the context of therapeutic proteins: An introductory overview. Post-translational Modification of Protein Biopharmaceuticals. 2009:1-14

[14] Kato A. Industrial applications of Maillard-type protein-polysaccharide conjugates. Food Science and Technology Research. 2002;**8**:93-199

[15] Spotti MJ, Martinez MJ, Pilosof AMR, Candioti M, Rubiolo AC, Carrara CR. Influence of Maillard conjugation on structural characteristics and rheological properties of whey protein/dextran systems. Food Hydrocolloids. 2014;**39**:223-230

[16] Karbasi M, Madadlou A. Interface-related attributes of the Maillard reaction-born glycoproteins. Critical Reviews in Food Science and Nutrition. 2018;**58**:1595-1603

[17] Tessier FJ. The Maillard reaction in the human body. The main discoveries

and factors that affect glycation. Pathologie et Biologie. 2010;**58**:214-219

[18] Billaud C, Adrian J. Louis-Camille Maillard, 1878-1936. Food Reviews International. 2003;**19**:345-374

[19] Taghavi F, Habibi-Rezaeid M, Amanie M, Sabourya AA, Moosavi-Movahedia AA. The status of glycation in protein aggregation. International Journal of Biological Macromolecules. 2017;**100**:67-74

[20] Hodge JE. Chemistry of browning reactions in model systems. Journal of Agricultural and Food Chemistry. 1953;**1**:928-943

[21] Ruiz-roca B, Navarro P, Seiquer I. Antioxidant properties and metal chelating activity of glucose-lysine heated mixtures: Relationships with mineral absorption across Caco-2 cell monolayers. Journal of Agricultural and Food Chemistry. 2008;**56**:9056-9063

[22] Ren GR, Zhao LJ, Sun Q, Xie HJ, Lei QF, Fang WJ. Explore the reaction mechanism of the Maillard reaction: A density functional theory study. Journal of Molecular Modeling. 2015;**21**:132

[23] ALjahdali N, Carbonero F. Impact of Maillard reaction products on nutrition and health: Current knowledge and need to understand their fate in the human digestive system. Critical Reviews in Food Science and Nutrition. 2019;**59**:474-487

[24] Delgado-Andrade C, Rufián-Henares JA, Morales FJ. Assessing the Generation and Bioactivity of Neo-Formed Compounds in Thermally Treated Foods. (1st ed.). Granada: Atrio SL. 2009. ISBN: 978-84-96101-76-0. (Chapter 1)

[25] Tamanna N, Mahmood N. Food processing and Maillard reaction products: Effect on human health and nutrition. International Journal of Food Science. 2015:1-6

[26] Ruan D, Wang H, Cheng F. The Maillard reaction. In: The Maillard Reaction in Food Chemistry. Cham: Springer; 2018. pp. 1-21

[27] Hodge JE. Origin of Flavor in Foods. Nonenzymatic Browning Reactions. Westport, CT: Avi Publishing; 1967. pp. 465-491

[28] Mendez DL, Jensen RA, McElroy LA, Pena JM, Esquerra RM. The effect of non-enzymatic glycation on the unfolding of human serum albumin. Archives of Biochemistry and Biophysics. 2005;**444**:92-99

[29] Van Boekel MAJS. Formation of flavour compounds in the Maillard reaction. Biotechnology Advances. 2006;**24**:230-233

[30] Parisi S, Ameen SM, Montalto S, Santangelo A. Chemical strategies against maillard reaction in foods. In: Maillard Reaction in Foods. Cham: Springer; 2019. pp. 33-43

[31] Cämmerer B, Jalyschko W, Kroh LW. Intact carbohydrate structures as part of the melanoidin skeleton. Journal of Agricultural and Food Chemistry. 2002;**50**:2083-2087

[32] Lund MN, Ray CA. Control of Maillard reactions in foods: Strategies and chemical mechanisms. Journal of Agricultural and Food Chemistry. 2017;**65**:4537-4552

[33] Sharon N. Nomenclature of glycoproteins, glycopeptides and peptidoglycans: Recommendations 1985. European Journal of Biochemistry. 1986;**159**:1-6

[34] Lis H, Sharon N. Protein glycosylation structural and functional aspects. European Journal of Biochemistry. 1993;**218**:1-27

[35] Freeze HH, Kinoshita T, Varki A. Chapter 46: Glycans in acquired human diseases. In: Varki A, Cummings RD, Esko JD, et al., editors. Essentials of Glycobiology [Internet]. 3rd ed. Cold Spring Harbor (NY): Cold Spring Harbor Laboratory Press; 2017. pp. 2015-2017. DOI: 10.1101/glycobiology.3e.046

[36] Varki A, Cummings RD, Esko JD, Freeze HH, Stanley P, Marth JD, et al. Symbol nomenclature for glycan representation. Proteomics. 2009;**9**:5398-5399

[37] Rabbani N, Thornalley PJ. Glycation research in amino acids: A place to call home. Amino Acids. 2012;**42**:1087-1096

[38] Van Nguyen C. Toxicity of the AGEs generated from the Maillard reaction: On the relationship of food-AGEs and biological-AGEs. Molecular Nutrition & Food Research. 2006;**50**:1140-1149

[39] Chen G, Smith JS. Determination of advanced glycation endproducts in cooked meat products. Food Chemistry. 2015;**168**:190-195

[40] Scheijen JL, Clevers E, Engelen L, Dagnelie PC, Brouns F, Stehouwer CD, et al. Analysis of advanced glycation endproducts in selected food items by ultra-performance liquid chromatography tandem mass spectrometry: Presentation of a dietary AGE database. Food Chemistry. 2016;**190**:1145-1150

[41] Matsui T, Nakamura N, Ojima A, Nishino Y, Yamagishi SI. Sulforaphane reduces advanced glycation end products (AGEs)-induced inflammation in endothelial cells and rat aorta. Nutrition, Metabolism, and Cardiovascular Diseases. 2016;**26**:797-807

[42] Sies H. Strategies of antioxidant defense. European Journal of Biochemistry. 1993;**215**:213-219

[43] Halliwell B, Gutteridge J. Oxygen toxicity, oxygen radicals, transition metals and disease. Biochemical Journal. 1984;**219**:1

[44] Boots AW, Haenen GRMM, Bast A. Health effects of quercetin: From antioxidant to nutraceutical. European Journal of Pharmacology. 2008;**585**:325-337

[45] Floegel A, Kim D-O, Chung S-J, Koo S, Chun O. Comparison of ABTS/DPPH essay to measure antioxidant capacity in popular antioxidant-rich US foods. Journal of Food Composition and Analysis. 2011;**24**:1043-1048

[46] Lobo V, Patil A, Phatak A, Chandra N. Free radicals, antioxidants and functional foods: Impact on human health. Pharmacognosy Reviews. 2010;**4**:118-126

[47] Migliore L, Coppedè F. Environmental-induced oxidative stress in neurodegenerative disorders and aging. Mutation Research, Genetic Toxicology and Environmental Mutagenesis. 2009;**674**:73-84

[48] Wu JQ, Kosten TR, Zhang XY. Free radicals, antioxidant defense systems, and schizophrenia. Progress in Neuro-Psychopharmacology and Biological Psychiatry. 2013;**46**:200-206

[49] Samaranayaka A, Li-Chan E. Food-derived peptidic antioxidants: A review of their production, assessment, and potential applications. Journal of Functional Foods. 2011;**3**:229-254

[50] Cheeseman KH, Slater TF. An introduction to free radical biochemistry. British Medical Bulletin. 1993;**49**(3):481-493

[51] Benjakul S, Lertittikul W, Bauer F. Antioxidant activity of Maillard reaction products from a porcine plasma protein-sugar model system. Food Chemistry. 2005;**93**:189-196

[52] Lertittikul W, Benjakul S, Tanaka M. Characteristics and

antioxidative activity of Maillard reaction products from a porcine plasma protein-glucose model system as influenced by pH. Food Chemistry. 2007;**100**:669-677

[53] Dong SY, Panya A, Zeng MY, Chen BC, McClements DJ, Decker EA. Characteristics and antioxidant activity of hydrolyzed β- lactoglobulin-glucose Maillard reaction products. Food Research International. 2012;**46**:55-61

[54] Stanic-Vucinic D, Prodic I, Apostolovic D, Nikolic M, Velickovic TC. Structure and antioxidant activity of β-lactoglobulin-glycoconjugates obtained by high-intensity-ultrasound-induced Maillard reaction in aqueous model systems under neutral conditions. Food Chemistry. 2013;**138**:590-599

[55] Jiang Z, Brodkorb A. Structure and antioxidant activity of Maillard reaction products from α-lactalbumin and β-lactoglobulin with ribose in an aqueous model system. Food Chemistry. 2012;**133**:960-968

[56] Gu FL, Kim JM, Abbas S, Zhang XM, Xia SQ, Chen ZX. Structure and antioxidant activity of high molecular weight Maillard reaction products from casein–glucose. Food Chemistry. 2010;**120**:505-511

[57] Jing H, Kitts DD. Chemical and biochemical properties of casein-sugar Maillard reaction products. Food and Chemical Toxicology. 2002;**40**:1007-1015

[58] Decourcelle N, Sabourin C, Aubry T, Guérard F. Emulsifying and antioxidant properties of a shrimp (*Pandalus borealis*) hydrolysate conjugated with xylose or dextran through the Maillard reaction by dry-heating in mild conditions. Journal of Food Research. 2014;**3**:144-157

[59] Jin H, Zhao Q, Feng H, Wang Y, Wang J, Liu Y, et al. Changes on the structural and physicochemical properties of conjugates prepared by the Maillard reaction of black bean protein isolates and glucose with ultrasound pretreatment. Polymers. 2019;**11**:848

[60] De Freitas J, Wintz H, Kim JH, Poynton H, Fox T, Vulpe C. Yeast, a model organism for iron and copper metabolism studies. Biometals. 2003;**16**:185-197

[61] Shulman A, Dwyer FP. Metal Chelates in Biological Systems In: Chelating agents and metal chelates. Dwyer FP, 1 editor. Nueva York, London: Academic Press Inc; 1964. p. 383-439

[62] Swaran F, Vidhu P. Chelation in metal intoxication. International Journal of Environmental Research and Public Health. 2010;7:2745-2788

[63] Irving HMNH, Williams R. The stability of transition-metal complexes. Journal of the Chemical Society (Resumed). 1953:3192-3210

[64] O'Brien J, Morrissey PA. Metal ion complexation by products of the Maillard reaction. Food Chemistry. 1996;**58**:17-21

[65] Sung WC, Chang YW, Chou YH, Hsiao HI. The functional properties of chitosan-glucose-asparagine Maillard reaction products and mitigation of acrylamide formation by chitosans. Food Chemistry. 2018;**243**:141-144

[66] You J, Luo Y, Shen H, Song Y. Effect of substrate ratios and temperatures on development of Maillard reaction and antioxidant activity of silver carp (*Hypophthalmichthys molitrix*) protein hydrolysate–glucose system. International Journal of Food Science and Technology. 2011;**46**:2467-2474

[67] Rufían-Henares J, De la Cueva SP. Antimicrobial activity of

coffee melanoidins—A study of their metal-chelating properties. Journal of Agricultural and Food Chemistry. 2009;**57**:432-438

[68] Wang H, Yang J, Yang M, Ji W. Antioxidant activity of Maillard reaction products from a yak casein-glucose model system. International Dairy Journal. 2019;**91**:55-63

[69] Grajek W, Olejnik A, Sip A. Probiotics, prebiotics and antioxidants as functional foods. Acta Biochimica Polonica. 2005;**52**:665-671

[70] Ames JM. Applications of the Maillard reaction in the food industry. Food Chemistry. 1998;**6**:431-439

[71] Corzo-Martínez M, Ávila M, Moreno FJ, Requena T, Villamiel M. Effect of milk protein glycation and gastrointestinal digestion on the growth of bifidobacteria and lactic acid bacteria. International Journal of Food Microbiology. 2012;**153**:420-427

[72] Einarsson H. The effect of time, temperature, pH and reactants on the formation of antibacterial compounds in the Maillard reaction. Lebensmittel-Wissenschaft and Technologie. 1987;**20**:56-58

[73] Hiramoto S, Itoh K, Shizuuchi S, Kawachi Y, Morishita Y, Nobuta Y, et al. Melanoidin, a food protein-derived advanced Maillard reaction product, suppresses *Helicobacter pylori in vitro* and *in vivo*. Helicobacter. 2004;**9**:429-435

[74] Ledesma-Osuna AI, Ramos-Clamont G, Guzman-Partida AM, Vazquez-Moreno L. Conjugates of bovine serum albumin with chitin oligosaccharides prepared through the Maillard reaction. Journal of Agricultural and Food Chemistry. 2010;**58**:12000-12005

[75] Echavarría AP, Pagán J, Ibarz A. Melanoidins formed by Maillard reaction in food and their biological activity. Food Engineering Reviews. 2012;**4**:203-223

[76] Chung YC, Kuo CL, Chen CC. Preparation and important functional properties of water-soluble chitosan produced through Maillard reaction. Bioresource Technology. 2005;**96**:1473-1482

[77] Riordan JF. Angiotensin-I-converting enzyme and its relatives. Genome Biology. 2003;**4**:225

[78] Takano T. Milk derived peptides and hypertension reduction. International Dairy Journal. 1998;**8**:375-381

[79] Aleixandre A, Miguel M, Muguerza B. Péptidos antihipertensivos derivados de proteínas de leche y huevo. Nutrición Hospitalaria. 2008;**23**:313-318

[80] Raia JJ Jr, Barone JA, Byerly WG, Lacy CR. Angiotensin-converting enzyme inhibitors: A comparative review. DICP. 1990;**24**:506-525

[81] Li GH, Le GW, Shi YH, Shrestha S. Angiotensin I–converting enzyme inhibitory peptides derived from food proteins and their physiological and pharmacological effects. Nutrition Research. 2004;**24**:469-486

[82] Hong X, Meng J, Lu RR. Improvement of ACE inhibitory activity of casein hydrolysate by Maillard reaction with xylose. Journal of the Science of Food and Agriculture. 2015;**95**:66-71

[83] Udenigwe CC, Mohan A. Mechanisms of food protein-derived antihypertensive peptides other than ACE inhibition. Journal of Functional Foods. 2014;**8**:45-52

[84] Jiang Z, Rai DK, O'Connor PM, Brodkorb A. Heat-induced Maillard reaction of the tripeptide IPP and

ribose: Structural characterization and implication on bioactivity. Food Research International. 2013a;**50**:266-274

[85] Jiang Z, Wang L, Wua W, Wang Y. Biological activities and physicochemical properties of Maillard reaction products in sugar–bovine casein peptide model systems. Food Chemistry. 2013;**141**:3837-3845

[86] Rufián-Henares JA, Morales FJ. Angiotensin-I converting enzyme inhibitory activity of coffee melanoidins Journal of Agricultural and Food Chemistry. 2007;**55**:1480-1485

[87] Jiang Z, Wang L, Che H, Tian B. Effects of temperature and pH on angiotensin-I-converting enzyme inhibitory activity and physicochemical properties of bovine casein peptide in aqueous Maillard reaction system. LWT—Food Science and Technology. 2014;**59**:35-42

[88] Hwang IG, Kim HY, Woo KS, Lee J, Jeong HS. Biological activities of Maillard reaction products (MRPs) in a sugar–amino acid model system. Food Chemistry. 2011;**126**:221-227

[89] Van Boekel M, Flogliano V, Pellagrino N, Stanton C, Scholz G, Lalljie S, et al. Review on the beneficial aspects of food processing. Molecular Nutrition Food Research. 2010;**54**:1215-1247

[90] Glösl S, Wagner KH, Draxler A, Kaniak M, Lichtenecker S, Sonnleitner A, et al. Genotoxicity and mutagenicity of melanoidins isolated from a roasted glucose–glycine model in human lymphocyte cultures, intestinal Caco-2 cells and in the Salmonella typhimurium strains TA98 and TA102 applying the AMES test. Food and Chemical Toxicology. 2004;**42**:1487-1495

[91] Taylor JL, Demyttenaere JC, Abbaspour Tehrani K, Olave CA, Regniers L, Verschaeve L, et al. Genotoxicity of melanoidin fractions derived from a standard glucose/glycine model. Journal of Agricultural and Food Chemistry. 2004;**52**:318-323

[92] Brands C, Alink GM, van Boekel MA, Jongen WM. Mutagenicity of heated sugar-casein systems: Effect of the Maillard reaction. Journal of Agricultural and Food Chemistry. 2000;**48**:2271-2275

[93] Wei Y, Chen L, Chen J, Ge L, He RQ. Rapid glycation with D-ribose induces globular amyloid-like aggregations of BSA with high cytotoxicity to SH-SY5Y cells. BMC Cell Biology. 2009;**10**:10

[94] Marko D, Habermeyer M, Kemeny M, Weyand U, Niederberger E, Frank O. Maillard reaction products modulating the growth of human tumor cells *in vitro*. Chemical Research in Toxicology. 2003;**16**:48-55

Chapter 12

Packaging Design Alternatives for Meat Products

Nícolas Mazzola and Claire I.G.L. Sarantopoulos

Abstract

This chapter connects the main requirements of meat product preservation with the most used plastic packaging structures, highlighting the role of packaging on the extension of food products shelf life and allowing their distribution and consumption in the most diverse areas of the world. An overview on the main degradation mechanisms of meat products is provided as background for the deeper discussions on flexible packaging films' compositions that is most used for meat packaging. Details on the performance of different sealing materials, gas barrier layers, and structural compositions as well as specific characteristics of packaging such as vacuum bags, shrink barrier bags, thermoforming, vacuum skin, and modified atmosphere packaging are discussed in this chapter serving as building blocks for an optimum packaging design, from food preservation to the final packaging end use. Finally, as part of an evolving world, new packaging trends are discussed, covering consumerism, sustainability, and functionality aspects.

Keywords: packaging, meat products, vacuum bags, shrink bags, skin packaging, modified atmosphere packaging, sealants, barrier resins, polyethylene, ionomers, polyamide, EVOH, PVDC

1. Introduction

The consumption of meat products refers back to the beginning of human history, with first ancient butcher activities believed to start around 10,000 BC [1]. Butchers are present in Bible parables and were an active community in ancient cities as Rome and London fostering the growth of pastoral economy around these places and supplying meat pieces to populations that could afford it. This arrangement was dominant till the eleventh century. The conservation and transportation of meat products were still scarce with rudimentary methodologies as salt and drying, limiting the industry to shops located into larger social agglomerates. Later on, during the 1700s the first slaughterhouse started to pack meat in the USA, fostering industry growth around cattle and pork, but still conservation and transportation were limited and challenging.

First ice trade companies were established in West Virginia, USA, during the early 1800s [2], and, with the volume expansion of ice commercialized during the late 1800s, the product was also a source of refrigeration for meat products across the USA and few other countries. With the advancements of refrigeration cooling systems, new containers and storehouses were designed to preserve perishable products and expand their movement among the countries, establishing logistic chains to transport chilled and frozen meat products safe and efficiently.

Another interesting finding during this era was the impact of CO_2 on meat preservation; it was found during meat product shipment from Australia to England that the use of solid CO_2 extended the shelf life more than meat held under ice [3]. At this time, rudimentary packaging materials were used to reduce contamination possibilities from production center to consumers.

During the 1950s, however, an important revolution started, with the decline of bulk sales based on the intermediation of butchers on sorting, selecting, and serving the consumers and the raising of sales model based on supermarkets, with products competing to each other for consumers' attention. In this new dynamic, packaging gained another important function besides protect the food: the ultimate communication with potential consumers [4].

Packaging is today a critical vehicle to provide physical protection for the meat products from environmental treats and control microbiologic growth during the shelf life period. On the other hand, packaging is a single variable in a complex equation of meat quality that comprises the characteristics of animals (age, nutrition, raising conditions), sanitary conditions of slaughtering facilities, storage, and transportation conditions.

With increasing demand for fresh, highly nutritional, and tasty food, the meat product production chain has invested on animal health technology programs as well as controlled transportation methods to extend product shelf life and enhance product quality at consumption moment. A proper packaging design can certainly contribute to microbial control and, when associated to proper storage conditions, can preserve good-quality fresh meat products up to 10 times versus unpackaged products, a key element to reduce food waste.

The global beef market today is formed by global players, focused on local market demands as well as food exportations. Looking at global beef production, half of world production is divided within the USA, Brazil, the European Union, and China, being the European Union and Brazil the two largest red meat exports in the world. Brazil has the largest commercial cattle herd in the world, adding 208 million heads, being almost 30 million slaughtered in 2016 [5].

2. Main deterioration mechanisms of meat products

The quality of meat production started with careful animal raising since prenatal and feeding during its life, forming a proper carcass, fat depositions, as well as number and size of muscular fibers. The overall concept of a high quality of meat products is based on a combination of several factors that can be correlated or individually analyzed. The main parameters used by industry can be described as (1) visual quality (color, muscle, and fat), (2) gustatory quality (texture, flavor, and odor), (3) nutritional quality, and (4) food safety [6]. The final product quality starts with proper handling of animals, primarily the food provided to them and living conditions, followed by adequate processing installations and controls, packaging, transportation, and storage. Several regulations and norms are applied in the different countries as best procedures to increase overall food quality and safety for consumers.

During storage meat products deteriorate because of pigment oxidation, oxidative rancidity, microbial growth, and even surface dehydration. Color is one of the key attributes evaluated by consumers during purchasing process, and it is determined by the state of the meat pigment, myoglobin, present into the meat surface. Fresh meat color changes to red, also known as "blooming," and is a derivate from the presence of oxymyoglobin (oxygenated state of myoglobin). In the absence of oxygen, the meat color turns into purple tone, with predominance of reduced myoglobin (deoxymyoglobin) on meat surface. Fresh meat products can also become brownish red with oxidation of myoglobin, generating metmyoglobin. This reaction occurs when small amounts of residual oxygen inside the vacuum-packaged fresh meat react with the pigment right after the vacuum packaging. The attractive red-pink color of cooked, cured meats is essentially an association of nitric oxide with native meat pigment that results in nitrosylhemochrome. This pigment is very sensitive to oxygen and light, so cured meats can fade during storage or under retail display if not protected by a proper package. Other inherent variables of beef products can affect color, such as age, gender, feed and pH, and microbial contamination. Storage conditions can also change the color of beef; especially when UV light is used at retailers, it can accelerate the conversion to metmyoglobin and change the color to brown tones. High-storage temperature also promotes discoloration.

Other quality parameters such as gustatory quality and food safety are strongly dependent on bacterial activities. The presence of microorganisms can lead to enzymatic deterioration and oxidation, but microbacterial growth is determined in several studies as the most important contributing factor for meat products' decay [6–8]. Vacuum-packaged meat products, in general, present extended shelf life compared to products in contact with air. There are two main bacteria that can proliferate into packaged meat products: *Pseudomonas*, which are dominant when enough oxygen is allowed into the package but suppressed by vacuum, and lactobacillus that can grow in vacuum environments. Both types are dominant and can outgrow each other under their favored conditions. The basic principle of meat packaging material design is to depress *Pseudomonas* growth and allow lactobacillus to proliferate, a genus of bacteria with a lower potential for deterioration. Proper permeability and gas composition around the product allowed by packaging materials combined with refrigerated conditions provide longer shelf life desired for beef products, reaching up to 120 days. Microbial spoilage may lead to changes in color and appearance as well as off-odor generation due to volatile metabolites.

The presence of oxygen can also lead to oxidation reactions of fats, accelerating odor changes and discoloration. Low O_2 level also contributes to avoid lipid, reducing the decay rate of meat quality and contributing to extended shelf life. The uses of packages with modified internal atmospheres (modified atmosphere packaging (MAP)) using intermediate to high levels of carbon dioxide (CO_2) combined with lower or high levels of oxygen (O_2) are alternatives to control bacterial growth and also preserve red tone of packaged meat. The most popular MAP gas mixtures are composed of blends of CO_2 and O_2, with O_2 being responsible to maintain red blooming tone of meat and CO_2 impacting the bacterial growth, limiting microbial deterioration. N_2 can also be added to the mix as an inert gas to balance internal and external partial pressures and reduction permeation rates of CO_2 and O_2. Mixtures of 80% of O_2 and 20% of CO_2 associated to storage temperatures in the range of 4°C can preserve meat up to 12 days and maintain its red color [7].

An alternative methodology to manipulate meat color by gas mixtures is to add carbon monoxide (CO) into MAP mixture of CO_2 and N_2, preserving meat red for more than 20 days. This application is controversial as it can mask spoiled products from consumers but mainly due to severe safety concerns with gas leak aging during packaging process and the inhalation of CO after opening the package. In the USA it is allowed to add CO at 0.4%, but this same approach is banned in other parts of the world [7, 9].

Due to the high water content on fresh meat cuts, unpacked products can present quick visual deterioration and weight loses. The liquid is also present into packaged meat cut, having deep impact on product appearance. Chilled temperatures can reduce the amount of exudate, but shrink bags and absorbent liquid pads can improve shelf appeal. The presence of liquids exudate can also be a factor that accelerates bacterial growth if environmental conditions allow, since it is rich in nutrients. This liquid fraction known as purge or drip is considered acceptable when found up to 2% of packaged product weight [6]; levels close to 4% are considered unacceptable and lead to economic loses.

In association with proper packaging processes, various preservation processes have been used for meat products over the years. Starting with drying and salting/curing, passing through chilled and frozen or heat treatment (from cooking till sterilization processes such retort), food preservation methods generally rely on suppression of key elements for deterioration (water, oxygen, light) and temperatures (cold and hot).

Additionally, food additives can be used to extend shelf life, as salts, sugar, liquid smoke, and spices and processes as smoking, etc. The uses of low temperature (−2 to −4°C) and freezing (below −15°C) are key methods for meat preservation used in the market due to the obtained reduction on the rate of microbiological growth. When water is frozen, all enzymatic processes and bacterial growth are reduced to a minimum, preserving meat quality for proper consumption in several cases beyond 12 months. The oxidative reactions, however, continue to affect meat quality with time through deteriorative mechanisms such as fat and pigment oxidation, which leads to color change. This discoloration increases especially when high concentration of purge or exudate from the meat is found which leads to pigment concentration on the meat surface.

Another major protein consumed by populations is the poultry, majorly distributed as whole frozen animals or frozen pieces. In several parts of the world, however, consumers are increasingly demanding for refrigerated poultry. One of the persistent problems of refrigerated poultry submitted to vacuum packaging is that when consumers open the package, they sometimes encounter what is referenced as "confinement odor," derivate mainly from lipid oxidation. In general, odor is the main limiting factor for consumers to accept the product as fresh; odor issues occur in general prior to the amount of bacteria achieve critical levels [10].

In summary, analyzing contributing elements for meat quality, the first one is the meat product itself (muscle and fat) and its initial microbial counts which can result on off-odor and gas production, discoloration, and changes on flavor. For processed meat products, the formulation is a key factor on shelf life, depending on ingredients and preservation method uses—e.g., cure, smoking, cooking, etc. The second parameter is the gas environment around the meat, associated with vacuum packages and MAP to control microbiological growth and chemical reactions. Light has also impact on color of meat due to interactions with pigments. The last critical parameter is the storage and transportation temperature which directly affects the decay rate, microbial growth, and dripping. Correct packaging design associated with good manufacturing practices and proper transportation and storage are key to deliver a high-quality product to consumers.

3. Packaging processes and requirements

There are several packaging possibilities available in the market to be used with meat products, comprising different materials such as metals, plastics, and carton. This chapter is fully dedicated to explore the different combinations for plastic packaging, a very versatile and efficient packaging presentation for such products used in formats as vacuum packages, modified atmosphere packaging, pasteurization, sterilization, freezing, and other nonthermal processing such as high pressure processing (HPP).

3.1 Flexible packaging components

Meat products are broadly packaged using flexible plastic packaging materials or a combination of flexible plastic packaging and rigid containers or carton boards. As packages are in general subjected to low storage and transportation temperatures, mechanical strength of polymeric materials is a must for proper protection of packaged goods in the final applications. Since first wraps are based on polyvinylidene chloride (PVDC) polymers till multilayer films with a combination of several attributes, meat packaging is one of the most complex fields into food packaging, combining a list of materials and processes to achieve proper protection, allowing consumption into the most remote areas of the planet.

Several material families are abundantly used in packaging structures, highlighting polyethylene (PE), polypropylene (PP), polyethylene terephthalate (PET), ethylene vinyl alcohol (EVOH), polyamides (PA), and PVDC as well as different copolymers within such polymer families.

Aiming to illustrate the main elements of meat packaging, the different features can be bundled into few elements that provide food protection: (1) sealing for proper hermeticity, (2) barrier to provide proper internal environment for food protection, and (3) abuse resistance to avoid failures due to mechanical impact and/or internal puncture from cured and bone meat pieces. The different elements are combined into single packaging structures via modern processes such as coextrusion, which multiple extrusion screws pump materials into a single extrusion die and combine the materials all together, adding up to 13 layers of different materials to deliver all needed packaging features. Another traditional methodology to combine flexible polymeric packaging materials is to adhere all together using lamination adhesives and/or extrusion coating. Although both processes can provide proper functionalities, the use of coextrusion has grown rapidly during the last years over lamination as it simplifies production process, eliminating one step and producing packaging films with all performance requirements directly from extrusion.

The role of packaging designers, from the material selection to the fabrication process selection, is key to maximize the packaging impact on extending product shelf life, reducing footprint from both packaging and product and providing correct end of life for recycling, reusing, or disposing.

3.1.1 Sealants

Sealants are the first key element on a proper packaging design, especially for vacuum and modified atmosphere packages. The most common sealing or welding methodology is based on heat sealing, either using constant heated sealing bars, induction sealing, or impulse sealers, always targeting for fastest possible sealing process to guarantee maximum productivity in a packaging line.

Equal to thermal processes, ultrasonic sealing also generates material melts to achieve molecular bonding of the layers. The major difference is that heat is generated internally in the packaging material itself by mechanical vibration rather than by conduction from the external layers to the inside sealing surfaces.

The selection of a sealing material for a flexible packaging structure depends on deep analysis of packaging usage requirements, filling process and machinery, possible product contaminations in the sealing area, and possible thermal processing. Hermetic sealing is obtained when the sealant material from both packaging faces can be untied and remains tied till consumption moment.

The parameters involved in a sealing process are basically the temperature of the interface to be melted, the melting temperature of the polymer, the polymer chain diffusion rate, melt strength, and crystallization rate. When melted and pressured, the two surfaces are in contact over a wide area and in fractions of seconds the diffusion of chains between the two sides occur, creating "molecular entanglements"—the chain diffusion is mainly due to Brownian movements of molecules and chains reptation [11]. After cooling, polymeric chains are recrystallized and hold the surfaces together. The main objective of sealing process is to create a hermetic bundle that prevents packaged content to leak, a factor that is imperative for vacuum and MAP formats. For packages that are subjected to thermal processing such as pasteurization or sterilization, besides hermeticity, the seal strength is another key variable to be considered during sealant material selection.

If mechanical stress is applied to separate the two surfaces while the material is still molten, the "entanglements" generated by chain diffusion generate a force called hot tack strength. This property is especially important for automatic packaging machine. If there is a minimum time of cooling and crystallization, the property is called heat seal strength—it can be measured after a certain time in a universal testing machine.

The main materials used as sealing layers or webs in flexible packaging are PE and its copolymers. PE was accidentally discovered in the early 1930s and in few decades becomes the largest volume polymer produced globally reaching a total close to 100 million metric tons per year. Based on ethylene polymer backbones and with wide range of comonomers technically available to be combined with ethylene, this polymer family presents a broad range of mechanical and thermal properties due to the combination of crystalline and amorphous proportions. PE molecular architectures can be composed of linear or branched molecular segments defined by the different polymerization and catalyst technologies available commercially. Traditional division of different polyethylene types by density is described in **Table 1**.

Packaging designers generally select PE resins at density ranges below 0.920 g/cm^3 as sealants materials, more preferably the ones produced using single-site catalysts (also known as "metallocene resins") due to their homogeneous composition and superior sealing properties. Metallocene polymers represent the newest class of PE made using molecular catalyst systems, with just one reactive metal center surrounded by molecular structures that hinders access of different monomers and comonomers during polymerization process. These molecular catalysts can control molecular parameters in higher degree when compared to the heterogeneous systems as Ziegler-Natta or chromium catalysts. The result is a very narrow distribution of comonomers across the polymer composition and sharp melting point.

The lower the density, the lower the heat-sealing temperature, meaning packaging can be sealed faster or using reduced energy amount. The heat seal initiation temperature (HSIT) has been obtained when the surface temperature reached an amount of amorphous material of 77% for PE [12]. Lower heat seal initiation temperature means that crystal domains of PE microstructures are small and with multiple defects, being easily molten to create a flowing material for heat-sealing

Resin family	Density range (g/cm^3)
High density polyethylene (HDPE)	0.941–0.965
Medium density polyethylene (MDPE)	0.925–0.941
Linear low density polyethylene (LLDPE/LDPE)	0.915–0.925
Very low density polyethylene (VLDPE)	0.900–0.915
Elastomers/plastomers	0.865–0.905

Table 1.
Classification of different polyethylene by density.

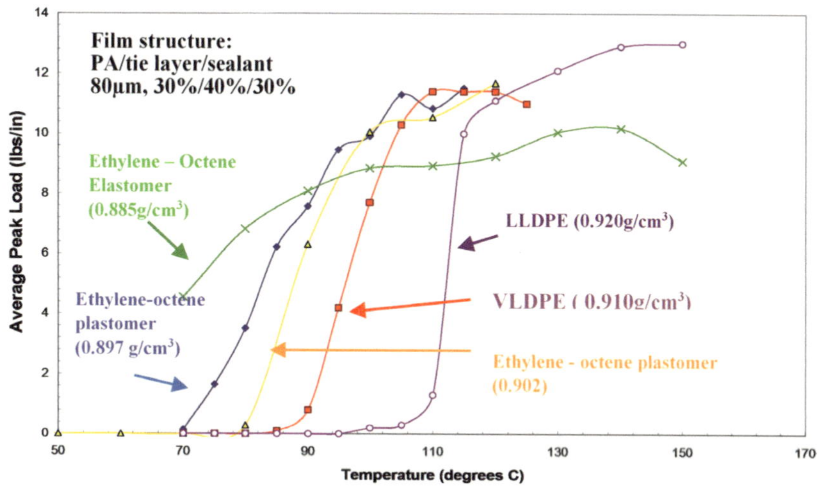

Figure 1.
Effect of polyethylene resin density into heat-sealing behavior of different sealing resins.

purposes. **Figure 1** demonstrates the effect of PE resin density into heat-sealing behavior of different sealing resins, highlighting the broad range of thermal properties a PE resin can achieve depending on the amount of comonomer incorporated to reduce density.

For meat packaging applications, it is very common to have contaminants such as grease, blood, and meat purge in the sealing area during packaging filling. In order to have proper sealing, materials with low-melting temperature are able to flow around contaminants or push them outside the sealing area to provide hermetic sealing. In the case of vacuum bags, the presence of folds and wrinkles is also very common, leading to hermeticity problems as small channels are formed in between the two sealed packaging webs. Sealant materials such as ethylene-based polyolefin plastomers (POP) and elastomers (POE) are best options to solve these problems and create hermetic packages due to their sharp melting point and low viscosity during heat-sealing processes. The ability to seal through contamination and close securely the microchannels is denominated in the industry as caulkability. Commercial brands of POP are Affinity™ from The Dow Chemical Company and Exact® from Exxon Mobil.

Sealants based on ethylene copolymers can also present different features when polar comonomers such as vinyl acetate, methacrylic acid, ions, etc. are added to the polymer composition; very particular properties can be obtained. Ethylene vinyl acetate (EVA) resins were used in blends with polyethylene materials few decades ago to reduce melting temperature of the sealant composition till more efficient plastomers were found and applied for meat packaging. Particularly in the case of shrink barrier bags, EVA copolymers can present a dual role, as they serve as sealants and also tie layers for the PVDC barrier layer. EVA polymers can be found in different suppliers such as the Dow Chemical Company with Elvax® and Braskem.

Within ethylene-based copolymers, analyzing specifically the meat packaging applications, ionomers are a class to be highlighted due to some unique features. Ionomers are based on a copolymer of ethylene with carboxylic acid and partially neutralized with ions, forming a polymeric structure with ionic bonds within its microstructure [13]. They were created and commercialized by Dupont since the 1960s under Surlyn®, with products based on Zn, Na, and K ions for different packaging and non-packaging applications. Such chemical structure creates a polymer with reduced crystallinity, enhanced clarity, superior sealing properties, especially

Figure 2.
Vacuum packages with polyethylene and ionomers as sealants and the combined effect of secondary seal and protein adhesion.

when sealing area is contaminated with liquids or fine powders, and strong interaction with different proteins present in the meat products. The intimate interaction of packaging and meat product [14, 15], particularly for Na- and Zn-based ionomers, is known as protein adhesion and used in many applications such as vacuum bags, thermoformed trays, and skin packaging to reduce the amount of purge liquids to be trapped in between the packaging material and the product, enhancing shelf appeal and extending shelf life. Another complementary feature found in ionomer sealant materials is the ability to generate strong adhesion between two film surfaces under vacuum when exposed to temperatures up to 90°C for few seconds. This adhesion is denominated by secondary seal and, associated with protein adhesion, creates a unique shelf appearance for vacuum packages as well as skin packages. **Figure 2** demonstrates two examples of vacuum packages, one with standard polyethylene sealant and another ionomer as sealants and the combined effect of secondary seal (water bath 85°C during 1 second) and protein adhesion.

In general sealed packages need tools as scissors or knives to be opened by consumers, but on the other hand, several packaging presentations are already commercialized with special sealant compositions that allow hermeticity to be maintained along all transportation and storage process but can be easily open by a consumer. There are two main mechanisms to create easy open seals: the first one is based on the delamination of sealing material from the adjacent layer, and the second one is obtained by selected polymer compositions with a certain level of incompatibility to generate cohesive failures when subjected to pulling stresses. For the delamination or burst peel, combinations of ionomers as sealants and HDPE as backing layer are a possible choice, and for cohesive failure, there are different formulated products such as Appeel® and Sealution™ from the Dow Chemical Company designed to be sealed into different materials such as PET, PP, PS, and PE and present easy open characteristics. Several other product combinations such as polybutylene (PB) and EVA, PB and LDPE, and ethylene acrylic acid (EAA) and LDPE can also be tailored to generate easy open properties.

A second product family to be highlighted as potential sealant layer for meat packaging is polypropylene and its copolymers. This class of sealants present higher melting point, from 140 to 161°C, and can stand sterilization processes that uses high-temperature exposure with no leakage. Packaging structures containing PP as sealant materials can be used in several packaging formats such as pouches, thermoformed trays, and vacuum bags.

3.1.2 Barrier resins

Once proper sealing is achieved in a given packaging, a second key element for meat packaging design is barrier layer. Different meat products require different gas and water vapor barrier protection. At the same time, different packaging materials can present the ability to modify the movement of gases and vapor molecules when different partial pressures of gases and vapors are found in either side of the packaging materials creating a flux from regions of high concentration to regions of low concentration. Permeability is defined by polymeric material ability to allow gases and vapor absorption on packaging surface and the transportation rate in the bulk of the polymer, being the first driver dependent on solubility parameters of permeant on polymer and the second on the free volume, crystallinity levels, and chain flexibility to allow permeant to move through polymer structures. **Figure 3** describes typical permeability ranges of main materials used for packaging films for two main permeants that can affect food quality, oxygen (O_2), and moisture.

For meat packaging applications, one of the pioneer materials used in such structures was the polyamides (PA), extensively applied on barrier packaging or as mechanical enhancer for heavy duty applications. Flexible packaging market recognized coextruded films with PA layers as structures with high tensile strength, improved thermoformability as well as enhanced impact and puncture resistance, being the sole election for applications such as bone-in meat.

PA resins are a large group of polymer structures, developed originally by DuPont in 1939. Polyamides are mostly aliphatic, linear polymers composed by the amide group, as repeating unit in the polymer chain, separated by hydrocarbon unit. Polyamides may be synthesized either by (A) polycondensation of divalent carboxylic acid and divalent amines or by (B) polycondensation of difunctional amino acids containing both one amine and one carboxylic acid functionality in the same molecule [16]. The amide groups are capable of forming strong electrostatic forces between the –NH and the –CO– units (hydrogen bonds), reducing free volume and therefore permeability, producing high melting temperatures, strength and stiffness, and chemical resistance. The increasing number of amides groups into the molecule increases the number of hydrogen bonds and reduces free volume, reducing gas permeability as well.

The most used PA resins for multilayer packaging structures are polyamides 6, 6.6, 11, and 12, the copolymers such as PA 6/6.6 and PA 6/12, and the terpolymers

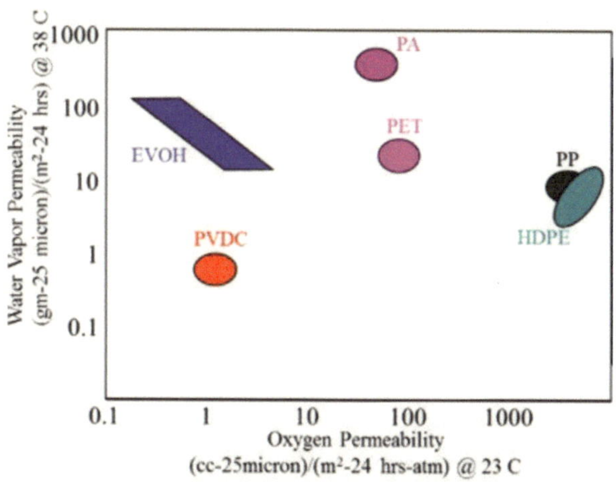

Figure 3.
Typical permeability ranges of barrier materials for food packaging.

as PA 6/6.6/12. The PA is used in general as core layer, surrounded by polyolefin tie layers and PE resins as external and sealing layer. The resulted packaging films combine hermetic sealing, improved toughness and medium O_2 barrier.

The amide units present strong interactions with water, motivating PA resins to absorb water from 2 to 20%. These water molecules are inserted into the hydrogen bonds, loosening the intermolecular attracting forces and increasing gas permeability through the polymer. The different PA compositions exhibit a melting temperature ranging from 178 to 260°C and different levels on moisture absorption—the higher the melting point, the higher the potential water absorption. The reduction on crystallinity is also found in copolymers and terpolymers, resulting on higher transparency, higher free shrinkage, improved thermoforming, and higher puncture resistance. As terpolymers are more permeable to gases, this material class can be advantageously used for certain meat products as vacuum-packaged salamis that demands a certain CO_2 permeability. Commercial producers of PA include UBE and BASF.

Another class of barrier polymer is ethylene vinyl alcohol (EVOH). First commercialized by Kuraray in 1972, it is a polar polymer widely used as oxygen barrier material for several applications in packaging industry, including meat packaging. As a semicrystalline copolymer of ethylene and vinyl alcohol, it also presents strong interactions among each polymer chains (inter- and intramolecular bonding) that leads to improved gas barrier performance, similarly to the mechanisms described for PA.

Along the years, EVOH has gained participation into packaging markets by replacing metallized films and foil due to its superior barrier maintenance when subjected to flex cracking. This barrier material presents good processability, and it is generally used as core layer into coextruded films in combination with polyolefin type of sealants and a tie layer based on a polyolefin modified with maleic anhydride. This is particularly relevant for EVOH due to its hydrophilic nature, with water molecules interacting with polar hydroxyl groups in EVOH when exposed to humid environments, leading to a plasticization effect that distance between adjacent molecules (free volume) particularly into the polymer amorphous phase [17]. The impact of moisture into EVOH barrier properties is well-known [17–19], and its performance as barrier material will be dependent on environmental conditions; the final package will be exposed.

Coextruded films are rapidly demonstrating high potential to reduce the impact of moisture into EVOH polymers, serving as an important tool for improved packaging design. Two approaches are generally used to protect EVOH from humid environments, the thicknesses of the layers protecting EVOH and the moisture barrier layers of such grades, being the most effective solution is the use of a material with higher moisture barrier in the side with highest relative humidity and a material with lower moisture barrier in the side with lower relative humidity [17, 19], concluding that it is not sufficient to pack EVOH layers in between thick high-moisture barrier layers, but the asymmetric design can deliver improved performance even with thinner layers. Combinations of EVOH with other polymers are also commonly found in the different packaging structures. Coextruded films based on EVOH, PA, and polyolefin polymers can present unique features as high barrier, improved mechanical resistance, and sealability to form hermetic barrier packages in thermoformed, vacuum bags, pouches, and other formats.

The level of oxygen barrier on EVOH resins is determined by the ratio of vinyl alcohol and ethylene fractions randomly distributed into polymer chains, determining factors for crystalline levels of EVOH resins. Higher levels of ethylene create more flexible polymer chains, resins with reduced melting point, crystallinity, and increased free volume, allowing more gas molecules to permeate through the system. EVOH resins with reduced ethylene contents present improved oxygen barrier properties, reaching levels lower than 1 $mL.m^{-2}.day^{-1}$ at commercially used thicknesses, but are more rigid and with higher melting point, resulting on packaging films with poorer mechanical performance. Considering recent advancements on extrusion controls for layer thicknesses and its variability, it is becoming more common to find commercial products with EVOH layers from 2 μm (vacuum bags) to 20 μm (thermoformed films), making this barrier resin a very competitive solution for high barrier applications.

A very particular barrier material is the chloride-based polymer-denominated PVDC. It is essentially a polyvinyl chloride (PVC) with a second chlorine atom per monomeric unit, representing a molecule with around 70% on weight of chlorine composition, creating a material with extremely reduced free volume that resulted in low permeability to gases. It was created during the 1960s by the Dow Chemical Company targeting to generate packaging films that extend shelf life of different food products through controlling oxygen and moisture permeability [20, 21], something unique as most of the polymer provides barrier for just one or another. PVDC homopolymers, however, due to their strong intermolecular forces, present melting temperatures from 198 to 205°C and degradation temperatures close to 210°C [22], reducing operational window available for proper extrusion of flexible packaging structures. In order to expand the extrusion processing boundaries, different comonomers such as vinyl chloride and methyl acrylate are added to the molecular structure and reduce melting point and expand processing window in 40–50°C [22], being the first type of comonomer mostly used in monolayer film applications for household wraps and sausage chubs and the second in coextruded films for fresh meat packages. All commercial PVDC resins are copolymers.

Coextruded films that use PVDC as barrier layer are in general composed of polyolefin external layers and EVA as tie layers. EVA polymers, generally with 9–18% VA contents, present enough polarity to provide proper adhesion to PVDC and polyolefin layers in a five-layer structure or decent sealing performance when just three layers are available. Combine PVDC polymers with other materials that demand higher processing temperatures such as PA and PET are quite challenging with traditional extrusion processes due to the low degradation temperature PVDC has, becoming usual to have PVDC barrier resins combined with low-melting temperature polyolefin grades as plastomers and ethylene copolymers such EVA.

When different barrier resin alternatives are considered for a meat packaging structures, one particular topic to assure proper protection along all product shelf life is the barrier maintenance when exposed to high-moisture environments. PVDC permeability properties are unaffected by relative humidity, delivering the

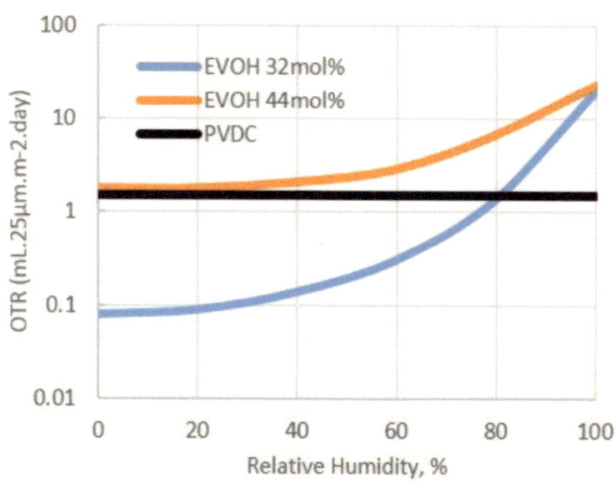

Figure 4.
Oxygen permeability for different barrier materials when exposed to several moisture levels.

same protection for dry and high-moisture environments. This particular feature was one of the main reasons PVDC is the dominant barrier technology for fresh meat packaging shrinkable films around the world, allowing the meat to reach shelf life levels up to 100 days.

Figure 4 demonstrates the impact of moisture on oxygen barrier properties of several typical barrier package films.

3.1.3 Abuse layers

Mechanical requirements in meat packaging applications are strong dependent on product characteristics, packaging processes and storage, and distribution environments. The type of meat product is the first element to be evaluated during the definition of packaging structures and thicknesses. Meat pieces with bone tips, cured skins, and sharp edges are the most critical class and demand packaging films with high puncture resistance. This property, however, can be divided into two different variables, puncture force and puncture deformation at break, and each of them is important for different aspects of packaging chain.

The combination of meat pieces with sharp edges and vacuum process can lead to complex issues due to vacuum loses after packaging. The origin of such issue is the vacuum process itself that projects the film on the meat surface and deforms the film on top of any existing irregularity with a very high deformation speed. When films are deformed at high speed, the time available for chains to rearrange and orient toward the applied tension is reduced, causing an early break. Lower deformation rates, in general, generate higher force responses from materials due to stress-hardening effects. Also, if packaging elongation is too high during vacuum process, the final film thickness at the tip of sharp meat piece is reduced and favors the packaging material to fail at this point with any subsequent deformation during transportation or handling. The dependence of deformation speeds on force is shown in **Figure 5** for two different coextruded films with 90 μm, the first composed of 40% PA 6 and 60% LLDPE and a second with 10% EVOH 38 and 90% LLDPE; the films with PA demand higher force to deform, while films with EVOH can deform more under lower force.

Another critical element to consider during material selection for a packaging with improved mechanical protection is the storage environment. As temperatures are reduced closer to polymer glass transition temperature, the failure mechanism of plastic packaging moves from ductile to fragile, reducing the capacity of plastic materials to absorb energy due to deformation under stress. These properties, however, can be manipulated by molecular design and polymer compositions that increases polymer flexibility and creates semicrystalline structures with improved toughness. An example of two different polyethylene resins with the same density and different dart energy responses at a range of temperatures are described in **Figure 6**.

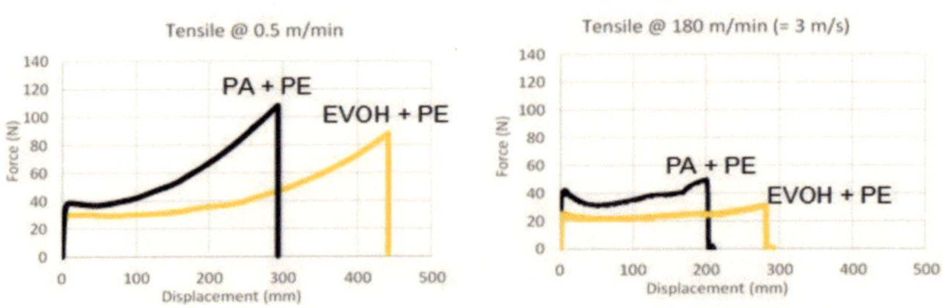

Figure 5.
Tensile force response of two packaging films (PA + PE and EVOH + PE) deformed at two different rates.

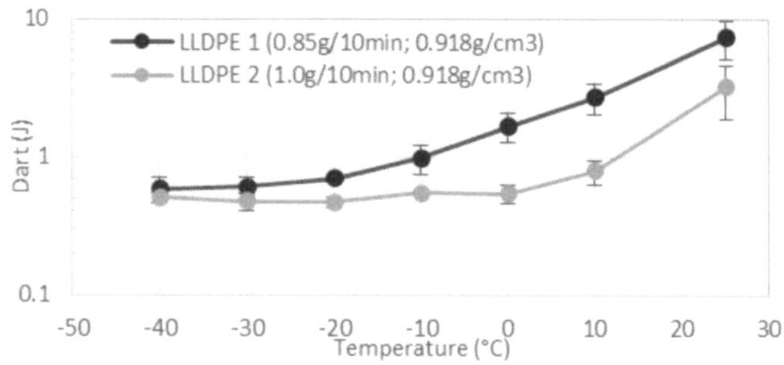

Figure 6.
Effect of environment temperatures on impact energy of PE resins.

The LLDPE 1 presents improved dart resistance vs LLDPE 2 for temperature ranges from −40 to 25°C, although the difference is smaller when temperatures are reduced to ranges toward PE glass transition temperature.

3.2 Packaging applications and formats

There are several studies [4, 7] that also indicate packaging as the main alternative for most of the brands to communicate with consumers at sales point; in Brazil, for example, 81% of purchase decisions are made in front of the shelves. About 10% of brands exposed in supermarkets have support from advertisement tools in television, magazines, and the Internet to promote their products; all the others rely on their own packaging to convince consumers to be the selected one during purchase process.

Packaging presentations and formats are chosen depending on packaging technologies that meat processors have access to purchase, can afford, and have proper machinery to use it besides all understanding on target consumers and market position. The same meat product can be packaged with simpler packages as film wraps to highly engineer packaging films with improved food protection and added consumer convenience through easy open, reclose, and improved shelf appeal.

One of the simplest packaging formats currently in use are non-barrier bags or plastic wraps, combined or not of trays or paper wraps, generally found at butcher stores or small supermarkets. Butcher shops still represent a traditional form of meat product purchase, and packaging used can protect meat against environmental contaminations, but no barrier to oxygen or moisture losses are expected from this type of packaging.

Another simple packaging presentation commonly used by retailers is the combination of trays made from polystyrene (PS), expanded polystyrene (EPS), PET, with PVC based wraps, or shrink films. For fresh beef, this combination has enough oxygen permeability (~60,000 $mL.m^{-2}.day^{-1}$ 23°C and 90% RH) to form oxymyoglobin on meat surface and maintain its red blooming color, an important factor on customer perception of fresh meat quality. PVC wrap can be manual or automatically applied, being heated at the bottom of the package to maintain the dimensions but with no hermetic sealing is achieved, allowing purge liquids to leak from package. As no barrier films are employed, packaging itself has minor influence on meat product shelf life. This type of packaging has an average thickness of 7 μm, the tray and film wrap weights from 15 to 20 g for a 1 kg portion, and fresh meat pieces packaged with this technology have an expected shelf life of 3–7 days when 1–4°C are used for storages. For poultry packages, there are some other shrink films based on PE that in combination with EPS trays are used as packaging of freshen and frozen poultry, but no barrier or hermeticity is obtained.

Vacuum barrier packaging types, on the other hand, restrict oxygen access to meat surface (OTR levels from ~5 to 20 $mL.m^{-2}.day^{-1}$ a 23°C and 90% RH), reducing the growth of spoilage bacteria such as *Pseudomonas* and favors the competing *Lactobacillus* species. Film structures can be based on EVOH and PA combinations with PE as sealants and possibly external materials with high thermal resistance such as PET—extrusion tie layers are needed to combine all distinctive layers. For refrigerated fresh meat, optimum storage condition is in the range of 1.5 ± 0.5°C [5] and residual oxygen levels as low as 0.15–0.20% [8]. On color, on the other hand, with the absence of oxygen, the formation of reduced myoglobin is favored, which results on a fresh meat with purple color tone. For processed meat products, the low OTR levels obtained with this packaging film help with color preservation and minimize fat oxidation, contributing to extend shelf life.

There are several types of vacuum package presentations used for meat products available in the industry, being the most common nonshrink vacuum bags, thermoformed packages, shrink barrier bags, and vacuum skin packaging.

Nonshrink barrier bags were one of the first presentations of meat packages available in the market, remaining today as a competitive alternative especially for processed meat goods such as cured pieces and sausages for wholesale. These packages are in general made from a preformed bag that receives the meat product, vacuum is applied till preset values, and subsequent heat sealing is performed generally using impulse sealing bars. These packaging films are based on coextruded structures with PA and/or EVOH as barrier layers, maleic anhydride-modified polyolefin as tie layers, and PE or ionomers sealant layers, adding from 5 to 9 distinctive layers with total thicknesses ranging from 70 to 150 μm depending on packaging size. From the packaging material requirements, polyethylene can provide hermetic heat sealing, even with grease or purge contamination, and ionomers deliver a combination of hermetic sealing and secondary sealing, eliminating meat exudate areas. Barrier layer types and thicknesses are selected according to gas permeability needs for each of the products. The packaging of selected bone-in meat products demands packaging films with enhanced puncture resistance and high-oxygen barrier, which are obtained by the combined use of EVOH and thick PA layers or EVOH and ethylene copolymers with improved puncture resistance such as ionomers, ULDPE, and metallocene LLDPE.

In order to improve shelf appeal and reduce packaging weight, a new packaging format with shrink properties to better shape packaged good was introduced into the market. The shrink barrier bag technology was originally developed by W.R. Grace (now Cryovac division of Sealed Air), and it is currently being practiced by other companies around the world. It consists on a biaxially oriented packaging structure produced via double or triple-bubble processes that is bottom sealed into a bag and receives the meat piece; after it, the vacuum is pulled, and the package is hermetically sealed using impulse sealing bars. The formed bags are submitted to contact with hot water or air (from 60 to 90°C) for few seconds and shrink forming a protective layer very close to the meat surface. Shrink barrier bags are the most efficient packaging solution when the ratio of the packaged product and packaging weight is considered; a 3 g shrink barrier bag can take the shelf life of a fresh meat piece up to 90 days when refrigerated at 0°C.

The film structures use majorly PVDC as barrier layers. A typical film structure takes also sealant layers of POP- and EVA-based tie layers adding 90% of film composition; PVDC barrier layer adds the remaining 10% of film weight. After extrusion, the films are in general subjected to crosslinking processes allowing proper dimensional stability during transportation and the possibility to overlap bags during sealing process without having one sticking to each other. Total thickness varies from 50 to 70 μm. The film can be printed, or labels can be inserted internally into the package.

Thermoformed trays are another type of packaging used to pack meat products. The packages are generally formed by the use of two film structures, one for the bottom part of the package, with 100–300 μm, and another for the lid, with 60–150 μm, adding a package weight close to 10 g for 1 kg package size. The thermoforming process to form the bottom of the package is based on three steps: (1) the bottom film is heated by a heated metal plate, (2) the film is vacuum formed into a tray, and (3) the bottom tray and lid film are sealed together after the package is filled with product. Alternatively to vacuum, a mixture of gases can be injected into the thermoformed packaging to create proper atmosphere for meat preservation. In this case the bottom film used to be a rigid or semirigid structure.

The film used for thermoformed bottom tray should have good thermoformability. This performance is related to the final thickness distribution along all tray profile after vacuum forming; thicker tray walls offer improved mechanical strength and reduced gas permeability to preserve food. Tray corners are the most critical spots, presenting, in general, final thickness after thermoforming reduced to 25% or more of the original film thickness. Combinations of PA and EVOH in coextruded films with polyethylene are in general the selected packaging structure for thermoformed bottom trays, as it combines excellent thermoformability and toughness of PA, with low oxygen permeability from EVOH and hermetic sealing from polyethylene. There are also EVOH grades with improved thermoformability, allowing deeper trays (more than 7 cm deep) to be formed with proper barrier maintenance along the tray profile.

MAP is used as an alternative to preserve fresh meat while maintaining its red blooming color, creating a positive combination of shelf life extension and improved shelf appeal. Structures used for MAP can be composed of flexible thermoformed trays and lids or rigid barrier thermoformed trays and flexible lids. Ideal designs for MAP are in general heavier (30–40 g) than simpler vacuum packaging structures due to the needed proportion of product and gas volume from 1:1 to 1:3, demanding higher head space. The complex barrier structure combined to higher weight increases total packaging cost when compared to vacuum bags, fact that drives MAP to be found in single or small premium portions that are ready to serve.

VSP are the latest vacuum packaging technology introduced for meat products. It consists on a rigid tray and a flexible plastic skin formed around the packaged good, creating a new possibility for the consumers to interact with a packaged meat piece. Due to premium appearance and general high cost, it is used to pack premium beef cuts and single pieces, being able to extend shelf life of fresh meat products from 21 to 35 days with a packaging weight of 13–19 g. VSP are growing very fast and expanding its reach to several other meat products, with strong penetration into cured meat pieces, with single pieces, premium presentations as well as poultry, following similar positioning.

During packaging process, the meat is placed on the rigid tray (premade or thermoformed in line), enters into a chamber where flexible film is being heated, and will be vacuum formed around the packaged good. The process itself is similar to thermoforming, but for VSP the formed films are heated at much higher temperatures, achieving values from 150 to 220°C to allow proper forming around most complex food shapes and formats. The packaging film needs to combine the required barrier to protect the meat piece, very high transparency to allow proper visualization of the product, puncture resistance for bone-in meat pieces, hermetic sealing to avoid contaminations, and easy open to facilitate consumer access to the product. Typically, the forming film structure is based on ionomer or easy open compositions as sealant materials, oxygen barrier layers, and overall high puncture resistance materials as ionomers, ethylene plastomers, or EVA. Films can also be subjected to crosslinking processes to improve their thermal resistance and formability around complex food products.

Format	Expected shelf life	Typical packaging weight
EPS tray and PVC wrap	5–7 days	17 g
Shrink barrier bags	90 days	3–5 g
Thermoformed trays	30–35 days	6–10 g
MAP	14–21 days	30–40 g
VSP	30–35 days	13–19 g

Table 2.
Different meat packaging formats existing into the market, expected shelf life, and typical weight of each format.

As a summary of commercially available fresh meat packaging technology, **Table 2** presents a scheme of packaging alternatives, expected shelf life, and typical weight of each format.

Retort packaging is a particular packaging structure that is designed to fulfill the requirements of retort preservation process and used to pack cooked meat and ready meals. The retort process subjects packaged products to temperatures from 115 to 150°C inside pressurized autoclaves to sterilize the food for periods ranging from minutes to few hours, resulting on a final product that is shelf stable at room temperatures for several years. Film structures for retort packages are in general laminated multilayer films with higher melting point sealant materials as polypropylene homopolymer, barrier layers composed by aluminum foil, SiO_x- or AlO_x-coated films or coextruded EVOH-based films, and laminated external printed layer based on PET. The packaging films used for retort packages should be thermal and chemically stable to stand the retort temperatures without the migration of chemical compounds into the packaged product above limits established by legislations.

Some typical structures for the different packaging formats are described in **Table A1** at Appendix. Selection of materials within each product family is critical to adequate the film performance to each of packaging machinery and operational conditions existing into the market as well as different products to be packed.

4. Innovative solutions for meat products

The packaging used for food products has several functions among its life; it has evolved to attend the demands and needs of the current social organization we have today. Packaging has evolved since the beginning of its creation, being today a highly complex piece of engineering that is targeted to preserve food with minimum amount of material to be economically viable and designed for minimum environmental impact [7]. The society, however, continues to demand new solutions that enhances life quality and solve new challenges, including accessible quality food.

Food protection and preservation are two basic features of every packaging technology introduced into the market, but in the recent years, a growing concern on packaging waste has been raised in the different regions of the world. Packaging design has a major impact on recyclability, leading manufactures to progressively frame their packaging with recyclability as a clear driver. The election of certain materials in detriment of others has a direct impact the expansion of different recycling chains.

Although PVDC containing packages are still dominant technology for fresh meat packaging around the world, with more than 500K MT of packages being produced annually, concerns with chlorine-based polymers end use and by-products formed during incineration are raising globally [18] and led the industry to find possible alternatives to produce shrink barrier bags with EVOH as alternative barrier technology. Europe was the pioneer continent to adopt such technology, led by developments on triple-bubble extrusion processes that combined EVOH as barrier technology with different materials for sealing, mechanical and moisture protection, and thermal resistance. Such process eliminated the need of irradiation step into oriented films as high thermal resistance materials can be used as external layers and allowed bags to be overlapped during sealing process without sticking to each other. EVOH-based shrink barrier bags accounts today for roughly 15% of shrink barrier bag market.

Shelf life maintenance in high-moisture environments was always a critical concern from meat producers to adopt EVOH-based shrink barrier bags. Recent study [15], comparing highly engineered multilayered shrink barrier bags based on EVOH barrier layers and different polyolefin resins and traditional PVDC based bags, was performed and has indicated both packaging solutions achieved similar meat quality levels, under controlled storage. Such indications and raising concerns with chloride-based polymers reinforce the potential growth of shrink barrier bag presentations based on EVOH barrier layers.

Poultry packaging industry, on the other hand, has created several alternatives to extend shelf life of refrigerated chicken using structures that possess agents to capture odor of vacuum-packaged chicken or use MAP to extend the product shelf life. The packages with odor absorption technologies can allow vacuum-packaged chicken to stand longer storage periods and be consumed with reduced odor perceived when consumers open the packages.

It is not just packaging materials and novel manufacturing processes are fulfilling the main demands from meat packaging industry, there are several technologies on design and functionalities that also contribute to enhance food protection and consumers experience [23]. After the consolidation of easy-to-open packaging presentations, the possibility to close it again is also spreading for several packaging types. The reclosability can be obtained through the use of zippers or other closure devices and more recently through the use of specially designed lids with pressure-sensitive adhesive compositions, such as M-Resins® from Bostik, in one of the internal film layers. The lid film is sealed on a tray, and when the lid is pulled, the film has adhesive or cohesive failure and exposes the sticky internal layer, creating a reclosable possibility that helps with food preservation during consumption period.

The new organization and size of families today also have created a new demand for food industry, reducing packaging sizes or creating packaging presentations with multiple small portions to be consumed along a period. The long commute journeys in big cities also increased the demand for food products that could be consumed on-the-go, based on smaller packages, generally barrier trays or retort pouches that contain cured meat, soups, stews, and baby food products. Convenience is another big demand from consumers today, targeting to food presentations that are convenient to consume and easy to prepare, without deprive demands of healthy and fresh food. In face of such demand, the industry has created options of entire recipes packaged using MAP for a quick finishing at home, ovenable packaging structures and cooked ingredients to help meal cooking process, without eliminating the individual cooker touch and fresh food impression [23].

Advancements on food sterilization technologies are creating novel meat products presentations and demanding new packaging solutions. Retort packaging is a traditional sterilization technology used for meat products, but new approaches are gradually being adopted due to different benefits such as longer shelf life, elimination/reduction of food preservatives, as well as nutritional and flavor preservation [24, 25]. High pressure processing (HPP) uses high pressure from the compression of a fluid medium such as water, ranging from 300 to 800 MPa, in general combined with certain temperature exposure (form to sterilize meat products. The process inactivates harmful and food spoilage microorganisms, being more effective on microorganisms with greater structural complexity—e.g., parasites are more susceptible than virus [25]. This process was adopted for some meat products such as cold slices and meat-based ready meals. As packaging exposed to this process are subjected to high pressure, they need to be able to accept up to 15% reduction in volume and recover its volume after pressure is released, maintaining hermeticity, strength, and barrier properties. Common packaging structures for this process are based on barrier trays and lid combinations and barrier vacuum-sealed pouches.

Other sterilization methodologies are being explored in certain applications, and there are few examples also for meat products. The use of microwave for pasteurization process ready meals has been explored to extend shelf life and create shelf stable food presentations, always combined with oxygen and moisture barrier packaging structures. Even newer approaches as the use of plasma and electric fields [24] are being explored to inactivate pathogenic microorganisms and improve food safety but still in early exploration phases.

Another major demand from modern consumer is the end use of packaging after it completes its main function of food protection. Although sharing partially the same comonomers, EVOH and polyethylene are not fully compatible for mechanical recycling purposes. The combination of such materials on multilayer packaging structures is, however, very common to achieve good balance of sealing, barrier, and toughness to several applications. Compatibilization technologies are available to be used during recycling process of these combined streams [26], but a new generation of compatibilizers from Dow under Retain™ product family claim to allow EVOH- and PA-based packaging films to be mechanically recycled with no extra compatibilizers added [27]. The correct amount of compatibilizer is added into discrete layers, and the material possesses enhanced dispersion capacity to create a homogeneous recycled material as well as allow the incorporations of such waste into existing PE recycling streams.

Meat product industry is continuously looking into novel ways to enhance their product offering toward healthier and safer products based on combinations of process and packaging designs. Packaging designers should be able to combine these demands into packaging structures that offer maximum food protection with appropriated end use after the completion of their function.

5. Conclusions

The first functionality of packaging is to physically protect food against external contamination and loses, from product production through the consumption in every consumer home, passing along all transportation chain and sales point exposure. If properly designed having in mind the specificity of meat product to be packaged and the main factors that drive its deterioration, packaging can also contribute to extend shelf life and enhance consumer experiences. Packaging design has a major impact on final end use, determining the attractiveness for recyclability, reuse, or final disposal of a packaging film.

Several elements should be added when designing a package, such as sealant properties and final needs on hermeticity, strength and access to product, gas barrier to control microbiological growth, and final mechanical resistance that will allow packaging to stand possible challenges during transportation and handling. Besides material selections, the production process is key to reach some desired features such as shrinkage and mechanical strength.

The plastic packaging structures used today are the result of a continuous optimization exercise performed by packaging and food chains in the last decades and continue to drive the whole industry toward the elimination of plastic waste and reduction of carbon footprint. The developments on meat processing and sanitary techniques also had a major contribution on providing safe and quality meat products to the population around the world.

Appendix

Packaging format	OTR level	Typical structure (from sealing to external layers)	Typical application requirements	Typical use
Vacuum bag	Medium	PE (POP)/tie/PA/tie/PE	• OTR from 20 to 50 $mL.m^{-2}.day^{-1}$ • High puncture resistance • Caulkability	Cured meat, sausages
Vacuum bag	High	PE (POP)/tie/PA/EVOH/PA/tie/PE	• OTR from 1 to 5 $mL.m^{-2}.day^{-1}$ • High puncture resistance • Caulkability	Pasteurized sausages, seasoned meat products
Shrink barrier bags	High	POP/EVA/PVDC/EVA/POP	• OTR from 8 to 20 $mL.m^{-2}.day^{-1}$ • >30% free shrinkage • Sealing through contamination	Fresh meat
Shrink barrier bags	High	POP/EVA/PVDC/EVA/IO	• OTR from 8 to 20 $mL.m^{-2}.day^{-1}$ • >30% free shrinkage • Protein adhesion	Fresh meat
Shrink wrap + EPS tray	Low	LLDPE + LDPE	• Shrinkage • Printing quality	Frozen chicken
Thermoformed bottoms	Mid	PE/tie/PA/tie/PA or PP	• OTR from 20 to 50 $mL.m^{-2}.day^{-1}$ • High puncture resistance	Frozen sausages Frozen meat
Thermoformed lids	Mid	PE/tie/PA/tie/PE//PET	• OTR from 20 to 50 $mL.m^{-2}.day^{-1}$	
Thermoformed bottoms	High	PE/tie/PA/EVOH/PA/tie/PA or PP	• OTR from 1 to 10 $mL.m^{-2}.day^{-1}$	Pasteurized sausages, fresh meat, seasoned meat, cold slices
Thermoformed lids	High	PE/tie/EVOH/tie/PE//PET	• OTR from 1 to 10 $mL.m^{-2}.day^{-1}$	
Rigid barrier trays	High	PE/tie/EVOH/tie/PET	• OTR from 1 to 10 $mL.m^{-2}.day^{-1}$	
VSP	High	PE/EVA/IO/tie/EVOH/tie/IO/EVA/Easy Open	• OTR from 5 to 15 mL.m–2.day–1 • High clarity, puncture, and formability • Easy open	Fresh meat, cured meat

Packaging format	OTR level	Typical structure (from sealing to external layers)	Typical application requirements	Typical use
Retort pouch	High	PP//Al//PET	• High-oxygen barrier • High-temperature resistance	Ready meals, cooked meat
Retort pouch	High	PP/tie/PA/EVOH/tie/PP//PET	• High-oxygen barrier • High-temperature resistance	

/—Coextruded layers, adhered using tie layers or inherent compatibility between adjacent polymers //—Laminated films, adhered using lamination process and adhesives

Table A1.
Typical structures for the different packaging formats.

Author details

Nícolas Mazzola[1*] and Claire I.G.L. Sarantopoulos[2]

1 The Dow Chemical Company, São Paulo, Brazil

2 Institute of Food Technology—Ital, Campinas, Brazil

*Address all correspondence to: nmazzola@dow.com

References

[1] Butchers Guild [Internet]. 2019. Available from: https://www.thebutchersguild.org/meat-story-an-incomplete-chronicle-of-butchery [Accessed: 02 June 2019]

[2] Weightman G. The Frozen Water Trade: How Ice from New England Lakes Kept the World Cool. 2nd ed. London: HarperCollins; 2003. pp. 90-98

[3] Farber JM editor. Principles of Modified-Atmosphere and Sous Vide Product Packaging. 2nd ed. Lancaster: Routledge; 2018. p. 54

[4] Mestriner F. Inovação na Embalagem. 1st ed. Sao Paulo: M. Books; 2018

[5] IndexMundi [Internet]. 2019. Available from https://www.indexmundi.com/agriculture/?country=br&commodity=cattle [Accessed: 02 June 2019]

[6] Robertson G. Food Packaging—Principles and Practice. 3rd ed. Boca Raton: CRC Press; 2013. pp. 445-460

[7] Furquim NR, editor. Beef, Production and Management Practices. 1st ed. Nova Science Publishers; 2018. pp. 1-25; 123-151; 175-225

[8] Zhou GH et al. Preservation technologies for fresh meat—A review. Meat Science. 2010;**86**:119-128

[9] Smith J. [Internet]. 2007. Available from: http://usatoday30.usatoday.com/money/industries/food/2007-10-30-kalsec-meat-carbon-monoxide_N.htm [Accessed: 02 June 2019]

[10] Arvanitoyannis I. Modified Atmosphere and Active Packaging Technologies. 1st ed. Boca Raton: CRC Press; 2012. p. 748

[11] Stehling FC, Meka P. Heat sealing of semicrystalline polymer films. II. Effect of melting distribution on heat-sealing behavior of polyolefins. Journal of Applied Polymer Science. 1993;**51**:105-119

[12] Mazzola N et al. Correlation between thermal behavior of a sealant and heat sealing of polyolefin films. Polymer Testing. 2012;**31**:870-875

[13] Reynolds JP. A Surlyn® ionomer as a self-healing and self-sensing composite [thesis]. Birmingham: University of Birmingham; 2011

[14] van den Broek, ATJ US 6,581,359 B1 Process and Polymer Films for Fresh Meat Packaging [Patent]. 2000

[15] Rodrigues JBM et al. Evaluation of the effectiveness of non-irradiated and chlorine-free packaging for fresh beef preservation. Meat Science. 2017;**125**:30-36

[16] Goetz W. Polyamide for flexible packaging film. In: PLACE Conference; May 12-14, 2003; Rome, Italy

[17] Jakobsen M, Risbo J. A simple model for the interaction between water vapor and oxygen transmission in multilayer barrier materials intended for food packaging applications. Packaging Technology and Science. 2008;**21**:207-216

[18] Khanah Mokwena K et al. Ethylene vinyl alcohol: A review of barrier properties for packaging shelf stable foods. Critical Reviews in Food Science and Nutrition. 2012;**52**:640-650

[19] Maes C et al. Recent updates on the barrier properties of ethylene vinyl alcohol copolymer (EVOH): A review. Polymer Reviews. 2018;**58**:209-246

[20] Valérie Renard V. Saran PVDC barrier concepts for demanding co-extrusion and lamination applications. In: 10th TAPPI European

PLACE Conference; May 22-25, 2005; Vienna, Austria

[21] Paisley K. PVDC, new developments new opportunities. In: TAPPI USA PLACE Conference; September 16-20, 2007; St Louis, USA

[22] Collins S et al. The thermal stability of some vinylidene chloride copolymers. Polymer Degradation and Stability. 1999;**66**:87-94

[23] Dantas F. Tendencias em embalagem: Conveniencia, identidade, premiumização e sesorialidade. In: Meat Plastic Packaging Seminar, May 26-27, 2015, Campinas, Brazil

[24] Stoica M. Et al non-thermal novel food processing technologies. An overview. Journal of Agroalimentary Proc and Tec. 2013;**19**:212-217

[25] Cristianini M. High pressure processing of food and packaging requirements. In: Meat Plastic Packaging Seminar, May 26-27, 2015, Campinas, Brazil

[26] Torradas, J.M. Enhancing Package Performance, Value and Sustainability with Modifiers. Dupont White Paper. 2011

[27] Packed with Performance [Internet]. 2014. Available from: http://msdssearch.dow.com/PublishedLiteratureDOWCOM/dh_093c/0901b8038093c374.pdf?filepath=packaging/pdfs/noreg/273-05901.pdf&fromPage=GetDoc [Accessed: 20 August 2019]